2024 Banff Pork Seminar

Advances in Pork Production Volume 35

Banff, Canada
9-11 January 2024

ISBN: 978-1-7138-9232-8

Printed from e-media with permission by:

Curran Associates, Inc.
57 Morehouse Lane
Red Hook, NY 12571

Some format issues inherent in the e-media version may also appear in this print version.

Printed with permission by Curran Associates, Inc. (2025)

For permission requests, please contact University of Alberta
at the address below.

University of Alberta
Department of Agricultural, Food and Nutritional Science
4-10 Agriculture/Forestry Centre, University of Alberta
Edmonton, AB, Canada T6G 2P5

Phone: (780) 492-3651
Email: pork@ualberta.ca

www.banffpork.ca

Additional copies of this publication are available from:

Curran Associates, Inc.
57 Morehouse Lane
Red Hook, NY 12571 USA
Phone: 845-758-0400
Fax: 845-758-2633
Email: curran@proceedings.com
Web: www.proceedings.com

Advances in Pork Production

Volume 35, 2024

banff pork seminar

Advances in Pork Production

Volume 35

Proceedings of the 2024 Banff Pork Seminar

Co-operating Agencies

Alberta Pork

Alberta Agriculture and Forestry

University of Alberta

Published by:
University of Alberta
Department of Agricultural, Food and Nutritional Science

ISSN: 1489-1395

Editor–Papers
Ashley Steeple
Editor–Abstracts
Ben Willing
Copyeditor
Ashley Steeple

Conference Coordinator
Ashley Steeple

For information please contact our office at:

Telephone: (780) 492-3651
Email: pork@ualberta.ca
Website: www.banffpork.ca

Mailing Address:
Banff Pork Seminar
c/o Department of Agricultural, Food and Nutritional Science
4-10 Agriculture/Forestry Centre, University of Alberta
Edmonton, AB, Canada T6G 2P5

Preface

December 2023 – "Meeting challenges and moving forwards."

Firstly, a very warm welcome to the 2024 Banff Pork Seminar! For over half a century, people from throughout the industry have been coming together from across North America and the world to the beautiful scenery of Banff, to network, to learn and to share ideas. Our theme this year is very reflective of our industry currently "Meeting challenges and moving forwards."

The advisory committee has worked hard to bring a line-up of speakers that will cover topics important to many areas across the pork value chain. These cover the economic picture, sustainability, new technologies and others. We again focused hard to bring speakers in the breakouts from across the world to present on subjects that matter to the day-to-day farm operations which will help them improve moving forwards. Please be sure to spend sometime networking in the tradeshow area and taking in the poster session.

As I reflect upon 2023, there have been many challenges for our industry with low hog prices, high feed costs and increased inflation just to name a few. But what I do see is a group of very resilient, dedicated, and focused people passionate about their industry who day in and day out are looking at every possible angle to keep our industry viable and strong for the long term, we should all be very proud of that.

This event would not be possible without the dedication and support of many throughout our industry. We want to extend a sincere thank-you to all our sponsors, without them this event wouldn't be possible. We want to thank all of you, our participants, for travelling to Banff to support the seminar.

A special thank you to the Advisory Committee, Ashley Steeple, and team and our two long time program directors, Ruurd Zijlstra and Michael Dyck for their continued dedication and support of the seminar.

The hog industry continues to evolve with better nutrition, genetics, technologies etc. I encourage all of you to spend some time reflecting on something you learn or discuss at the seminar that you can take back to your area of the industry to continue to move us forwards in 2024.

Lastly thanks again for all of you taking the time to attend, enjoy the seminar and be sure to spend time connecting and enjoying some time in Banff.

Sincerely,
Steve M Davies
Chair, 2024 Banff Pork Seminar Advisory Committee

2024 BPS Advisory Committee

Chair

Steve Davies, Maple leaf Foods., Carman, MB

Program Directors

Ruurd Zijlstra, Professor, Swine and Carbohydrate Nutrition, Dept. of Agricultural, Food and Nutritional Science, University of Alberta, Edmonton, AB

Michael Dyck, Associate Professor, Reproductive Physiology/Biotechnology, Dept. of Agricultural, Food and Nutritional Science, University of Alberta, Edmonton, AB

University of Alberta

Ben Willing, Assistant Professor and Canada Research Chair in Microbiology of Nutrigenomics, Dept. of Agricultural, Food and Nutritional Science, University of Alberta, Edmonton, AB

Alberta Agriculture and Forestry

Jennifer Hannesson, Livestock Traceability Section, Edmonton, AB

Alberta Pork

Javier Bahamon, Quality Assurance Coordinator, Edmonton, AB

Industry Representatives

Maaike Campbell, Birnam Pork, Warwick Township, Ont.
Lindsay Case, PIC, Montreal, Que.
Melissa Da Costa, Baseline Pork, Harley, Ont.
Jen Demare, Swine Health Professionals, Steinbach, Man.
Andrew Fenton, DNA Genetics, London, Ont.
Samuel Lefebvre, Jyga Technologies, Saint-Lambert-de-Lauzo, Que.
Murray Pettitt, Prairie Swine Centre, Saskatoon, Sask.
Rick Prejet, Porcherie Lac du Onze, Notre Dame de Lourdes, Man.
Cordell Young, Precision Veterinary Services, Lethbridge, Alta.
Zhenbin Zhang, Cargill Animal Health & Nutrition, London, Ont.

Conference Coordinator

Ashley Steeple, University of Alberta, Edmonton, AB

We extend sincere thanks to the following companies and groups who have provided financial assistance to the 2024 Banff Pork Seminar and to the publication of these proceedings.

Sustaining Sponsors

AcuFast
Alberta Pork
PIC
Topigs Norsvin Canada

Premier Plus Sponsors

AgriHub Inc.
Dept. of Agricultural, Food and Nutritional Science, University of Alberta
Maple Leaf Foods
Olymel
RDAR
Steve's Livestock Transport
Swine Veterinary Partners

Premier Sponsors

Alberta Animal Nutrition Lectureship
Canadian Pork Council
DNA Genetics
dsm-firmenich
HyLife Ltd.
Jefo Nutrition Inc.
PigGen Canada
Robinson Bioproducts
South West Vets

... Sponsors of the 2024 Banff Pork Seminar (continued)

Select Sponsors

ACO FUNKI A/S
Agribution Canada Ltd.
Animal Health Canada, Emergency Management Division
APC
Asserva SAS
Automated Production
Barkman Concrete
Boehringer Ingelheim Animal Health Canada Inc.
Cargill
Carlo Genetics
DanBred Ltd.
Dechra Veterinary Products
EastGen
Farm Credit Canada
Grand Manufacturing
Grand Valley Fortifiers
Hams Marketing
Hipra Animal Health
Innovative Heating Technologies
Jyga Technologies
Kane Veterinary Supplies Ltd.
Kemin Animal Nutrition & Health
Ketchum Manufacturing Inc.
Merck
Minitube Canada
Modern Veterinary Therapeutics
Neogen
Novus International Inc.
Polar Pork Farms
Prairie Hog Country
Prevail Disinfectants (Virox Technologies)

... *Sponsors of the 2024 Banff Pork Seminar* *(continued)*

SEC Repro
SKOV A/S
Summit Technologies Inc.
Trouw Nutrition
Vereijken Hooijer
VIDO
Warman Veterinary Services
Zoetis

Supporting Sponsors

Bio Agri Mix
Hamlet Protein Inc.
JBS Live Pork
Le Porc Show
Maximus
PigCHAMP Inc. Farms.com
Rusk Medical Corp.

Index of Advertisers

... *Index of Advertisers* *(continued)*

Program of the 2024 Banff Pork Seminar
January 9-11, 2024

Plenary Sessions

Wednesday

- **Economics & Market Outlooks**

  Joseph Kerns, *Partners for Production Agriculture*

- **The Canadian Pork Industry: An Integrated Processor's View**

  Arnold Drung, *Conestoga Meats*

Thursday

- **Food Sustainability**

  Sylvain Charlebois*, Dalhousie University*

- **Lessons I Learned the Hard Way About Carbon**

  Marty Seymour, *Carbon RX*

- **Leading Through Crisis - Supporting Best Practices**

  Jason Shaw*, Winnipeg Fire Paramedic Service*

Morning Breakout Sessions

1. **Feed & Nutrition**

- **Food & Fuel Industry Co-Products: Considerations for Swine Nutrition Programs**

  Jay Squire

- **Amino Acids as Functional Nutrients for Pig Health**

  Dan Columbus, *Prairie Swine Centre*

2. **Piglet Management**

   - **Nutrition & Management of Weaned Pigs Fed Low Zinc Diets**

     Mike Tokach, *Kansas State University*

   - **Weaning Age Impact on Piglet & Sow Performance – Is Higher Always Better?**

     Luc Dufresne, *Demeter Veterinary Services*

3. **Technology**

   - **Point of Care Using Real Time Data in Sow Barns**

     Matthew Rooda, *SwineTech*

   - **ProHog Management – A Refresh on Margin**

     Steve Dziver, *Commodity Professionals*

4. **Innovators**

   - **Disease resilience indicators based on immune stimulation of blood from young healthy pigs**

     Vishesh Bhatia, *Iowa State University*

   - **Effects of beta-glucans on the intestinal inflammation in weanling pigs**

     Marley Conceicao dos Santos, *University of Saskatchewan*

   - **Burlap & buddies: Enrichment options to benefit piglets around weaning**

     Ashlyn Scott, *University of Manitoba*

   - **Biofilm formation, and Amoxicillin resistance of E. coli Isolated from wild and conventional pigs**

     Joshua Theriault, *University of Alberta*

- **FX Aherne Award for Innovative Pork Production**

  MealMeter 2.0 by *Dave Klocke*, *PigEasy*

Afternoon Breakout Sessions

5. Genomics

- **What is Genomic Selections & What Has Been its Impact on Current Genetics**

  Tom Rathje, *DNA Genetics*

- **Individual SNP Effects & Interesting Applications in Pig Breeding**

  Egbert Knol, *Topigs Norsvin*

- **Indicators & Genetics of Disease Resilience**

  Jack Dekkers, *Iowa State University*

6. Grow-Finish Management

- **Feeding & Management of Grow-Finish Pigs**

  Dan Bussieres, *Groupe Ceres*

- **Taking Care of Basics**

  Ken Engle, *Prairie Swine Centre*

- **Effect of Space Allowance & Marketing Strategy on Production Performance of Grower-Finisher Pigs**

  Willem Steyn, *SwiNE Nutrition*

7. Sow Management

- **Where Have All the Gilts Gone? The Impact of Mortality & Removals on Gilt Retention**

  Jayne Jackson, *PigCHAMP*

- **Nutrition & Feeding Behaviour of High Producing Sows in the Age of Global Warming**

  Bruno Silva, *Universidade Federal de Minas Gerais*

- **Prop 12 Sow Management – Lessons Learned**

  PJ Corns, *JBS Live Pork*

8. Swine Health

- **Internal Biosecurity – Is There Value For You?**

  Brad Chappell, *Topigs Norsvin*

- **Update on Sapovirus – An Industry Perspective**

  Matt Ackerman, *Pork Vet Solutions*

- **Streptpcoccus zooepidemicus – Impact on the Hog Industry**

  Richard Rusk, *Rusk Medical Corporation*

- **Blisters, Exports & Foreign Animal Disease Reporting – The Unexpected Impacts of Seneca Valley Virus**

  Glen Duzier, *Manitoba Agriculture*

Table of Contents

3. Technology

4. Innovators

5. Genomics

AcuFast™
The best way to predict the future is to Create it
(306) 667-1420 | www.acufastswine.com

TN70 MAKE YOUR BUSINESS A #1 HIT
THANK YOU TN70 FANS!
No extra labor needed, The TN70 nurses her own piglets
No 1. sow in the world
Brazil
Netherlands
Germany
Belgium
UK
Ireland
France
Russia
Hungary
Romania
Serbia
Czech Repu-
blic
Croatia
Austria
Slovakia
Poland
Lithuania
Greece
Spain
Argentina
Uruguay
Costa Rica
Colombia
Italy
Canada
USA
Mexico
China
Japan
Philippines
Vietnam
Norway
Sweden
Finland
Denmark
Bolivia
Ecuador
Peru
Paraguay
South Africa
Ukraine
Portugal
tn70.topigsnorsvin.com
Topigs Norsvin

ALBERTA PORK
Our mission is to deliver services so our producers and partners receive the best possible value.
The Alberta pork industry is measured on how we address the 4-Ps of sustainability:
PIGS
PEOPLE
PLANET
PROGRESS

The Canadian Pork Industry – An Integrated Pork Processor's View

Arnold Drung

President, Conestoga Meat Packers Ltd., adrung@conestogameats.com , 519-648-2505 (ext 236)

1. Introduction

The Canadian pork industry has been through a tumultuous 25 years since the ominous year of 1998. Until recently, most hog producers remember as the worst time for hog farming in living memory. Since then, the industry has faced dramatic currency swings, major disease issues, food safety crises and geo-political situations affecting exports. Throughout, the industry has been able to adapt and grow, maintaining its position as one of the Top 5 global exporters of pork. Many industry participants believe that we are now at an inflection point – does the industry continue on its past growth trajectory, move to a steady state or decline?

As an integrated, farmer-owned pork processor, Conestoga Meats has a unique view on the Canadian pork industry, spanning much of the value chain. In its view, the industry has a tremendous amount of upside, but it will be up to all industry participants to create the conditions that allow Canadian pork to truly take advantage of the opportunity.

2. Past is Prologue

If we look at the past 25 years, the industry has had to face some serious challenges while also having tremendous opportunities. Throughout this time, market forces and the capabilities of industry players drove adaptation to and exploitation of the situations that arose. The live hog price collapse of 1998 led first to an industry shake-out with non-competitive players leaving the industry. In the U.S., it led to significant consolidation and vertical integration, particularly as Smithfield bought large farming operations. From a Canadian perspective, the early 2000s saw the Canadian dollar drifting lower, and ultimately to an all-time low in early 2002. This insulated the industry to some extent on a global basis as costs were lower while export prices were higher. The industry grew during this time period but that came to an abrupt halt with the run-up in the CAD, taking it to an all-time high in late 2007. From an industry perspective, costs went up while export selling prices declined, a brutal set of circumstances that saw another industry shake-out. Marginal participants were forced out of the industry while stronger players were able to survive and further consolidate. We saw these players adapt to the conditions and leverage things such as crop and feed diversification in order to reduce costs. While the currency rise was rapid and high, it did force those remaining in the business to become better and ultimately strengthened the industry.

The same can be said of the disease outbreaks that took place over this time. From circovirus to PRRS to PED, the industry had to learn how to deal with each and was largely successful in each case. The inappropriately named Swine Flu

(H1N1) proved to be more challenging as it had nothing to do with the industry yet did have significant impact on it. It coincided with the difficulties of the 2008-2009 financial crisis and the recession that resulted from it.

Geo-political factors have also influenced the Canadian industry. After losing real access to the EU in the 1990s, there were high hopes with the opening of China and Russia in the 2000s. With access and the resulting business, companies became more aware of the need for stable relations with both countries. Unfortunately, the fact that they have governments that do not work on the same basis as Canada's soon manifested itself in suspensions, border issues and other problems that don't occur with countries governed similarly to ours. Access to Russia by major western exporters was ultimately cut off following Russia's Crimean invasion. Exports to China have been subject to suspensions both at a company and country level, largely based on the political climate between the two countries.

As a result, Canadian pork processors have worked diligently to further diversify exports, helped by the addition of new trade agreements such as the CP-TPP (Comprehensive and Progressive Agreement for Trans-Pacific Partnership). This agreement solidifies exports to Japan but also opens up key developing pork markets like Vietnam. The industry has been disappointed with the CETA agreement between Canada and the EU as it has not resulted in meaningful access due to non-tariff barriers.

With each of these challenges and headwinds, the key to survival and success has been a solid base business, and the ability to recognize the issues and rapidly adapt to the situation. This will continue to be critical as we move forward in the coming decades.

3. What the Industry Needs to Flourish in the Future

The formula for future growth of the Canadian pork industry still relies on long-standing fundamentals. These include competitive hog producers, competitive processing operations, a competitive trade environment and a competitive regulatory regimen.

3.1 Competitive Hog Producers

The basis for the success of any hog industry is the competitiveness of its hog producers. There is no globally competitive pork exporter that can survive long term without its hog producers having one of the best hog raising costs. Canada has been blessed with an abundance of land and natural resources which have made our country into an agricultural powerhouse. As we know, different parts of the country are amenable to different agricultural products. We can say that Canada is a great place to raise hogs but it is better to say that regions of Canada are a great place to raise hogs. Currently, three of the most competitive hog-growing regions in the World are Brazil, Iowa and Southern Ontario. We need to recognize this fact and plan investments and support accordingly. Our governments have made some significant “bets” on who will be the winners in various industries and backed them accordingly. This has been less of the case for the pork industry. With appropriate environmental and business regulations along with business risk management programs, Canadian hog growing regions can produce competitively. It is important that all members of the supply chain work to maintain that competitive position in order to ensure benefit for all.

3.2 Competitive Processing Operations

When looking at the global pork industry, it is obvious that certain players have invested heavily to drive down the cost of processing while increasing production capabilities/flexibility, and these tend to be the global leaders. In the global market, the low-cost producers set the pricing bar for all. Canada has had a mixed experience in this arena. While there are some globally competitive processing operations, many are not at a scale that allows for them to effectively compete outside of Canada. As we look to the future, Canadian companies must keep a close eye on their global competition and determine what it will take to truly compete against them. It will be up to individual competitors to determine their appetite for investment and risk in taking on global companies. There is no reason why scale sized Canadian processors can't effectively and profitably compete against the best in the world. We have the access to technology and the educated workforce that can achieve this.

3.3 Competitive Trade Environment

In order to be the most competitive, Canadian pork processors must have access to the largest number of pork importing countries possible. This allows for diversification of risk along with the ability to sell different pork products to countries with different requirements (and there are very diverse tastes and needs globally). Canada has been successful in signing trade agreements with many of the world's countries. However, the agreements have not always been in the best interests of pork or other agricultural commodities. If the government wants the industry to continue to grow and flourish, it must ensure that agreements are only entered into with jurisdictions that accept Canada's very high food safety standards as equivalent to their domestic programs. Under the CETA agreement, Canadian pork has been denied meaningful access to European markets due to non-tariff barriers that are hardly defensible. At time of writing, Canadian red meat associations have been attempting to block Canada's support of the UK accession to the CP-TPP due to the fact that the UK will not grant equivalence to the Canadian food safety system. It is incumbent on the government to ensure that Canada's agricultural industries have full access in trade agreements.

An opportunity to assist Canadian exporters is for the government to provide some form of risk protection for geo-political situations where the decisions of the government result in trade disruptions (e.g. China's ban on Canadian pork exports following the arrest of the Huawei executive in 2018-19). Companies experienced significant losses on products prepared and shipped but not delivered when this suspension was put in place. A government sponsored insurance program for these sorts of cases would be of tremendous assistance.

3.4 Competitive Regulatory Regimen

Canada has a globally recognized food safety regulator in the Canadian Food Inspection Agency. Our access to foreign markets is driven by the CFIA's ability to maintain regular market access with all of our trading partners. As we move into the future, it will be important for CFIA and its federal funder to invest in the technology and personnel required to enhance its cost of delivering inspection services while also providing the necessary support at embassies in global markets that accept Canadian pork.

A key component of the industry's ongoing success is the value of the Canadian dollar relative to other currencies, particularly the USD. As mentioned earlier, past currency shocks have been very challenging for the industry and it does make it easier when we have a lower CAD. However, in the spirit of what was written earlier, working to keep industry costs as low as possible sets it up with the best chance for ongoing success.

Conestoga Meats

Conestoga Meats is the processing division of Canada's only farmer owned, new generation co-operative, Progressive Pork Producers Co-operative Inc. The Co-op was founded 30 years ago in 1994 and purchased the Conestoga Meat Packers business in 2001. Since then, the company has grown tremendously and is now one of the five largest pork processing operations in Canada. Conestoga's experience of the last 20 years has mirrored that of the Canadian pork processing sector. However, the operation has had the added benefit of a land-based ownership group that has been able to weather many of the negative conditions outlined earlier in this document while also taking advantage of the positive opportunities that have arisen over time. A key strength of the Conestoga model is the integrated decision making that looks at all aspects of the value chain. When issues or opportunities arise, there is dialogue between the farm segment and the processing operation to determine the best course of action for the company as a whole. There have been times where it was determined that not going ahead with a farm change or a processing change was actually more financially beneficial to the organization as a whole and the change was not undertaken. An example of this was the choice to not proceed with the adoption of Ractopamine in hog diets. Conestoga identified early on that there were significant market opportunities available by not using the feed ingredient and these outweighed the potential farm benefit. It proved to be a positive decision from the outset.

When looking at the Canadian pork industry, a mindset of what benefits the industry as a whole will be a key driver of future success. The old win/lose model that has often existed between farmers and processors is out of date and when we look at successful global exporters such as Spain and Brazil, there is much more aligned decision making based on the concept of win/win.

4. Summary

The Canadian pork industry has been a success story, weathering numerous challenges over the past quarter century. As we look to the future, the World will need more protein (current UN-FAO estimates show a shortage of the beef, pork and poultry relative to projected increases in demand over the next 10 years). Canada is well-positioned to supply the World's increasing protein needs, particularly for pork. By being sharply focused on maintaining global competitiveness in farms, processors, trade agreements and regulatory regimens, the industry can continue to thrive and be a strong contributor to the economic success of Canada.

SVP Swine Veterinary Partners
Swine Veterinary Partners (SVP) is a group that provides a full range of animal health services to pork producers. Four veterinary clinics and pharmacies are established across the country and offer professional services as well as sharing their knowledge and expertise for the benefit of the Canadian swine industry.
HERD HEALTH
RESEARCH & DEVELOPMENT
YOUR HEALTH FILE ONLINE !
LABORATORY
PHARMACY SERVICES
DELIVERY SERVICES
TOGETHER
WE IMPROVE THE FUTURE
PRECISION Veterinary Services Inc.
PREMIERSHP Your leaders in swine health
DEMETER Veterinary Services Inc.
DEMETER Services Vétérinaires Inc.
ALBERTA
precisionvet.ca
1-403-406-3550
MANITOBA
premiershp.ca
1-204-233-7690
ONTARIO
demetersv.com
1-226-242-2575
QUEBEC
demetersv.com
1-877-847-5411

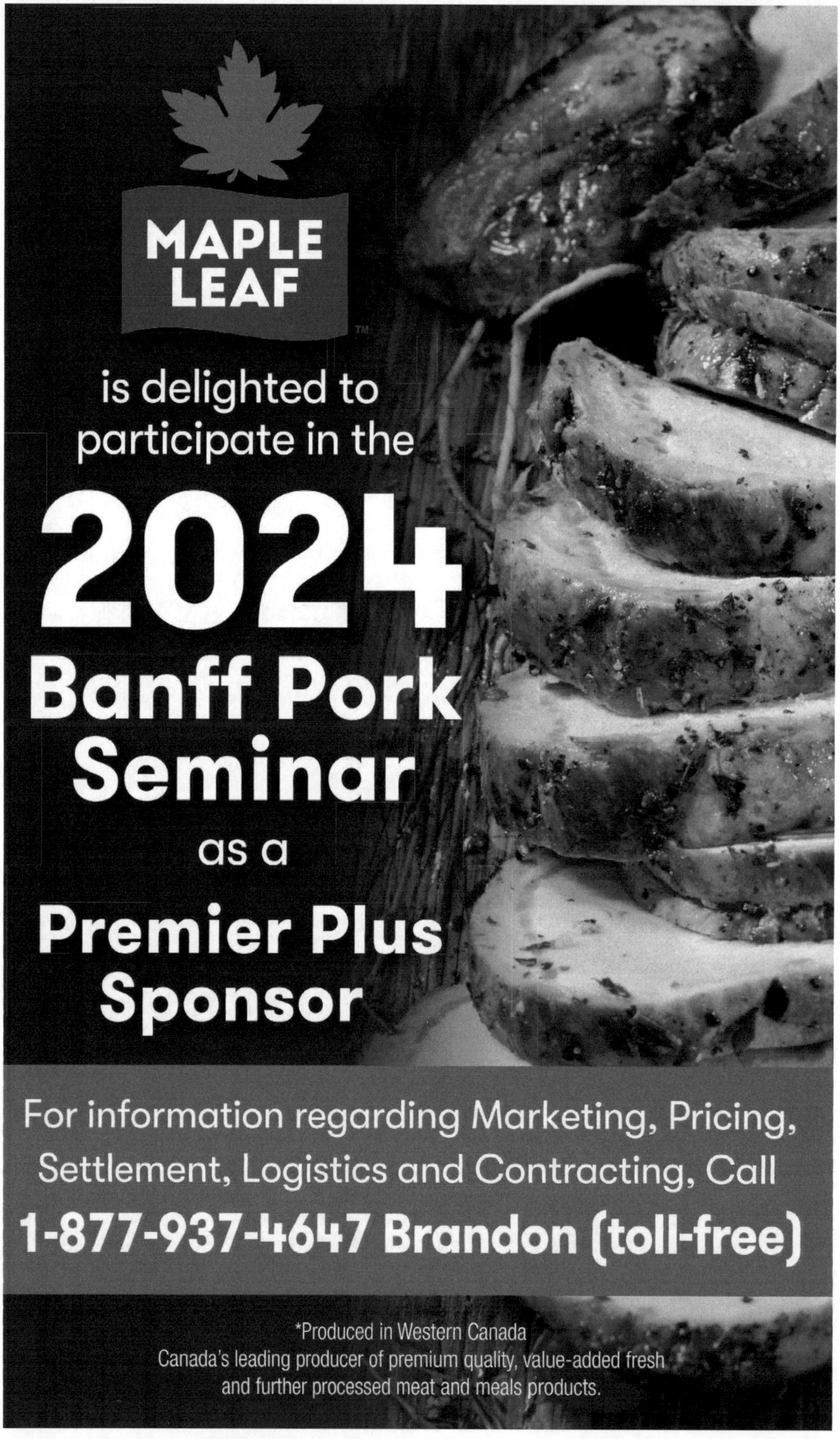
MAPLE
LEAF
is delighted to
participate in the
2024
Banff Pork
Seminar
as a
Premier Plus
Sponsor
For information regarding Marketing, Pricing,
Settlement, Logistics and Contracting, Call
1-877-937-4647 Brandon (toll-free)
*Produced in Western Canada
Canada's leading producer of premium quality, value-added fresh
and further processed meat and meals products.

Food & Fuel Industry Co-Products: Considerations for Swine Nutrition Programs

Jay Squire, M.Sc.

33516 Mitchell Line RR3 Lucan, ON, Canada N0M 2J0
jaymichaelsquire@gmail.com

1. Introduction

These past three years have been a reminder how vulnerable our swine industry is to volatile feed ingredient markets and supply chain issues. With continued margin pressure for producers, it is prudent to explore alternative feed ingredients as more cost-effective energy and protein sources for pigs.

Pigs have a unique advantage over other livestock and poultry in their ability to convert co-products from the human food and fuel industries into lean muscle protein. Co-products fed to swine can include ingredients like whey from the dairy industry, condensed distiller solubles from ethanol plants, bakery meal from bread recyclers and yeast from breweries. In the past, these types of ingredients had been referred to as by-products – suggesting an unwanted "waste product". As swine producers worked with their nutritionists to address food safety concerns and understand true nutritional value of these products it was important to begin referring to them as co-products to avoid the negative connotation of feeding waste to swine.

In Canada, there was rapid adoption of liquid feeding technology by Ontario swine producers during the 1990's and early 2000's. In general, this was a strategy to multi-phase feed and include fermented high moisture corn into swine diets to avoid costly grain drying. It was soon determined that Ontario has a strong human food processing industry and there were many liquid co-products that could be sourced at low cost (rather than dumped) and incorporated into the diets of swine.

There are many considerations when investigating co-products for swine feeding. This paper/presentation intends to address a few.

2. Dry Matter and Handleability

A particular co-product can look attractive on a dry matter basis but consideration must be given to how easily and effectively it can be included into swine rations. While one can assume that liquid products can be managed with liquid feed systems, it is possible to obtain ingredients that have dry matter levels too low for effective feeding. Depending on what's already included in the diet, a meaningful inclusion of the wrong co-product could create a mixture so wet that physical feed intake becomes a limiting factor to the pig consuming enough nutrients for maximum growth. There are also ingredients that can separate in storage and require regular physical agitation to ensure a consistent product is being pulled into the mix. Other, drier, ingredients can be fibrous or high in fat content which limits their ability to flow out of conventional ingredient bins. Again, physical agitation or "live bottom" floor auger bins may be required.

3. Nutrient Content & Digestibility

At first glance, a particular co-product can appear higher in value than what it may be. It's important to recognize that through processing, an ingredient may test high in protein, for instance, but the quality of that protein and its availability to swine may be lower. For example, a product like liquid corn steepwater from the corn wet milling industry has a crude protein 90-95% that of soybean meal but when it comes to lysine (an important first-limiting amino acid in swine diets), it has approximately half the amount. Further, the digestibility of lysine within corn protein is typically lower than soy products and as such, the amount of digestible lysine within steepwater may be closer to just one third that of the swine feed staple, soybean meal.

4. Safety and Stability

Food chain safety, for both humans and swine, is extremely important when evaluating co-products for feeding. It is common for the food and fuel industries to use cleaning processes between production batches and runs. It's important that assurances are made that co-products supplied to livestock producers are free from contamination. Another important factor to consider with swine feeding safety includes evaluating nutrient levels. Many food industry co-products can contain high levels of sodium that must be managed carefully to avoid toxicity. Pigs can also be quite susceptible to antinutritional factors and mycotoxins that can be concentrated in some co-product supplies. Lastly, shelf-life of co-products must be evaluated as a higher moisture content may make an ingredient more susceptible to fermentation or spoilage.

5. Cost & Availability

Once all the above considerations have been explored, the co-product in question can then be balanced against associated costs and its availability. Quite often, transport costs to farm are the largest expense associated with higher moisture co-products. For example, some dairy products can be as low as 4-5% dry matter. Transporting product that is 95-96% water to farm can eliminate any savings associated with the dry matter inclusion. Producers must also remember that there are hidden costs associated with animal performance due to higher potential for nutrient variability, utilities due to handling/storage costs and even some barn deterioration over time with lower pH and higher sodium products being used. It's also important to understand the availability associated with a particular product. Some food and fuel processors only have product available sporadically. Swine producers and nutritionists need to consider the impact on feed intake due to changes in ration palatability as well as the cost of having storage available for an ingredient not always being fed.

6. Conclusions

The Canadian swine industry will likely face ongoing margin pressures in future years. Producers should continue to explore options for lowering cost of production and this could include the incorporation of food and fuel industry co-products in swine feed. While not discussed here, another consideration and sustainability argument for ongoing co-product feeding to swine may be the importance of recycling carbon generated by the food and fuel industries into nutritious and delicious pork.

NEW
OVN®
OPTIMUM VITAMIN NUTRITION
UPDATED OVN OPTIMUM VITAMIN NUTRITION® GUIDELINES
Check and adjust vitamin levels for peak swine health and performance
GET THE UPDATED GUIDELINES
ANIMAL NUTRITION AND HEALTH
ESSENTIAL PRODUCTS
PERFORMANCE SOLUTIONS BIOMIN
PRECISION SERVICES
DSM

Amino acids as Functional Nutrients for Pig Health

Daniel A. Columbus

Prairie Swine Centre, Inc.; Saskatoon, SK, Canada S7H 5N9
Phone: 306-667-7432
dan.columbus@usask.ca

Introduction

Currently, diets are formulated to optimize growth performance and profitability of pork production, however, with societal and legislative pressure to reduce antibiotic use in livestock agriculture, including in swine, there is an increased need to consider how diet impacts animal health as well as performance. This includes an evaluation of nutrient requirements during disease challenge in order to formulate diets that support proper gastrointestinal development and function and immune response during disease challenge in addition to growth performance. Pigs are continuously exposed to microbial pathogens and immune-stimulatory antigens that negatively impact animal productivity, with the impact on immune status likely a major factor in the observed gap between performance potential and observed performance under commercial conditions. This decrease in performance can have a substantial impact on profitability of producers. The continued sustainability of pork production will necessarily involve a re-evaluation of nutrient requirements and adjustment of diet formulation.

Impact of immune challenge on performance and nutrient utilization

The reduction in performance observed during disease challenge has largely been attributed to the observed drop in feed intake, however, recent meta-analyses have demonstrated that both reduction in nutrient supply (i.e., feed intake) and nutrient utilization (i.e., maintenance requirements) both contribute to reduced growth. The proportion with which feed intake and nutrient utilization impact performance is dependent on the specific immune challenge, with enteric bacterial infections having a greater impact on nutrient utilization while the negative effects of mycotoxins and respiratory disease are more due to decreased feed intake (Pastorelli et al., 2012; Rodrigues et al., 2021a).

In general, immune stimulation alters protein and amino acid metabolism and utilization, with amino acids redirected from growth towards supporting the immune response. This limits amino acid availability for growth which is further exacerbated by reduced feed intake. As a result, amino acid supply for the immune response is partially met through a catabolism of muscle protein. However, the amino acid profile of muscle protein differs significantly from that of protein involved in the immune response (Reeds et al., 1994). Overall, while it is true that immune challenge will result in a reduction in feed intake and, therefore, reduce growth and nutrient requirements for growth, it is likely that amino acid requirements (i.e., amino acid ratios to lysine) need to be adjusted during times of immune challenge based on determination of requirements in pigs during immune stimulation.

Functional amino acids and pig health

Based on the amino acid profile of immune system components, glutamine, arginine, threonine, and the aromatic and sulfur amino acids would appear to play key roles in the immune response (Reeds and Jahoor, 2001). Therefore, supplementation with these amino acids during immune challenge, above the requirements for growth during, may be of benefit in mitigating the negative effects of immune challenge. Indeed, previous studies have demonstrated an increased requirement for growth for methionine and cysteine (Rakhshandeh et al., 2010; Litvak et al., 2013), threonine (Jayaraman et al., 2015; Wellington et al., 2018; McGilvray et al., 2019), and tryptophan (De Ridder et al., 2012) in response to immune stimulation. It is important to note, however, that amino acid requirements to maximize growth, with or without immune challenge, may differ from those required to maximize other outcomes, such as components of the immune response. This was demonstrated in a recent meta-analysis where it was demonstrated that dietary amino acid content above those for maximum growth result in improvements in key physiological responses, such as milk yield, litter size, immune response, and intestinal permeability (Ramirez-Camba and Levesque, 2023). So while we tend to base our diet decisions on growth performance, and the majority of immune stimulation requirements studies focus on growth, this may not be an appropriate outcome if optimizing pig health is the goal. The term 'functional amino acids' has been used to describe those amino acids supplemented in the diet for their roles beyond those for protein synthesis (i.e., lean gain). In the context of health, these include amino acids with significant roles in gastrointestinal health (e.g., barrier function) and immune response (e.g., antioxidant balance, acute phase response).

As indicated above, while past research has indicated alterations in nutrient metabolism, specifically of amino acids, the effect of supplementation of amino acids for pig robustness has only recently received significant attention. Supplementation of individual functional amino acids has been shown to improve growth performance of pigs under immune challenge. Wellington et al. (2018) demonstrated an increase in the threonine requirement in immune stimulated pigs and then further demonstrated (Wellington et al., 2019) that threonine supplementation resulted in improved growth performance of growing pigs fed high fibre diets during Salmonella challenge, most likely through support of improved gut barrier function (i.e., mucin production). Likewise, Koo et al. (2020) observed that increasing threonine content resulted in improved gut integrity (i.e., goblet cell density, tight junction gene expression) in pigs with induced intestinal inflammation as a result of being fed a ‘simple’ diet. Koo et al. (2021) also demonstrated improved immune status in pigs housed in unsanitary conditions when fed valine-supplemented diets. An improvement in growth performance was also observed by Jayaraman et al. (2015) and Trevisi et al. (2015) in pigs housed in unsanitary conditions or challenged with E. coli, respectively, when fed diets containing supplemental threonine and Le Floc’h et al. (2009) observed improved growth performance in weaned pigs provided supplemental tryptophan when housed in unsanitary conditions.

More recent work has examined the use of a blend of functional amino acids on growth performance and immune status in weaned and growing pigs. For example, Rodrigues et al. (2021b) fed grower pigs diets containing either a standard amino acid profile (NRC, 2012) or one in which threonine, methionine, and tryptophan were supplemented at 120% of the requirements for growth. Pigs fed the amino acid blend had improved growth performance and immune status (e.g., increased serum albumin and glutathione, reduced serum haptoglobin and superoxide dismutase) when challenged with Salmonella than their unsupplemented counterparts. The same amino acid profile improved growth performance and immune status (i.e., acute phase response, antioxidant balance) in Salmonella-challenged nursery pigs (Rodrigues et al., 2021b). Likewise, Valini et al. (2023), using the same amino acid blend, observed improved growth performance and immune status, including reduced body temperature and Salmonella shedding, and improved average daily gain, body weight, and nutrient utilization in grower pigs that were challenged with Salmonella and housed in unsanitary conditions. Rodrigues et al. (2021b) also observed decreased fecal myeloperoxidase, an indication of reduced intestinal damage, with the amino acid blend, indicating a role of these amino acids in supporting gut health during an enteric pathogen challenge. The effectiveness of supplementation with this amino acid blend may be dependent on specific conditions, as van der Meer et al. (2016) showed improved immune status but limited effect on growth performance in wean to finish pigs under low sanitary conditions even though the same amino acids were supplemented at the same level as Rodrigues et al. (2021b).

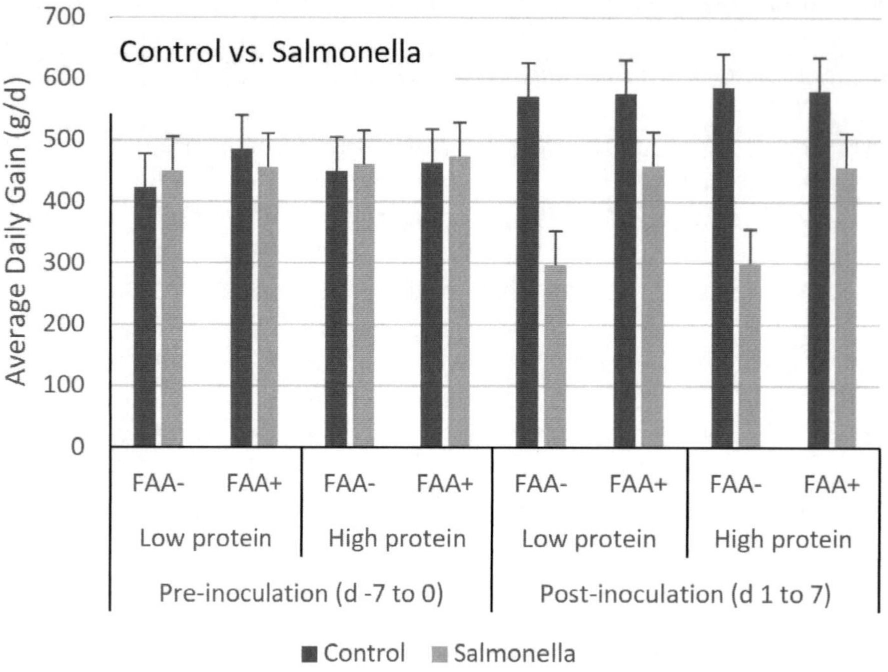

Figure 1. Growth performance of weaned pigs pre- and post-inoculation with saline (Control) or Salmonella and fed diets containing low or high protein (16.6 vs. 19.5%) with (FAA+) or without (FAA-) a functional amino acid blend supplying methionine, threonine, and tryptophan at 120% of NRC (2012) requirements for growth (n=8 pigs/treatment) Adapted from Rodrigues et al. (2021b).

Factors affected functional amino acid impact

The timing and duration of functional amino acid provision may impact the effectiveness of supplementation at improving performance and immune status. Rodrigues et al. (2021c) demonstrated a further increase in growth performance in pigs provided the blend of functional amino acids when the adaptation period prior to Salmonella infection was increased from 0 to 2 weeks (Fig. 2). Not all components of the immune response responded the same way to functional amino acids supplementation time, however, with acute phase response responding positively to a longer adaptation time while antioxidant balance did not. While there was no overall effect of amino acid supplementation, van der Meer et al. (2016)

observed improved average daily gain in the nursery period and improved feed efficiency in the finisher period in pigs housed in unsanitary conditions. Moreover, the response to sanitary conditions was greater during the nursery period, indicating that supplementation may be more beneficial in the immediate post-weaned period. This is supported by Rodrigues et al. (2021a) who demonstrated that the negative impacts of disease challenge are greater in the nursery period than in grower-finisher pigs, potentially due to a more immature intestinal tract in this period. Schweer et al. (2019) and Jasper et al. (2020) were able to improve growth performance in pigs inoculated with PRRS virus by increasing lysine:energy ratio at the time of challenge, however, no improvement was observed if the adjustment was made post-challenge (Gabler, 2021).

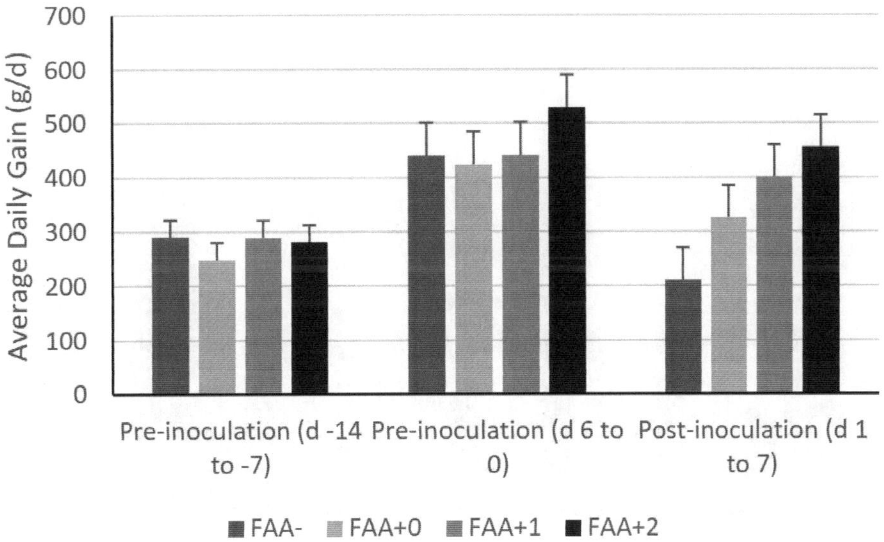

Figure 2. Growth performance of weaned pig pre- and post-inoculation with Salmonella and fed diets without (FAA-) or with a functional amino acid blend supplying methionine, threonine, and tryptophan at 120% of NRC (2012) requirements for growth during the post-inoculation period (FAA+0), for 1 week pre- and post-inoculation (FAA+1), or for 2 weeks pre- and 1 week post-inoculation (FAA+2) (n=8 pigs/treatment). Adapted from Rodrigues et al. (2021c).

Functional amino acid supplementation in the nursery period may also provide long-term benefits, as Rodrigues et al. (2022a) demonstrated improved growth in pigs that had received a supplemental functional amino acid blend in the nursery period but were fed a common grower diet at the time of Salmonella-challenged in unsanitary conditions. This blend of amino acids, however, was only partially able to attenuate the negative effect of disease challenge in low-birth weight pigs during the subsequent disease challenge. Other components in the diet may also affect functional amino acid effectiveness, as supplementation with the same blend of functional amino acids was able to improve performance of nursery pigs that had received a plant-based diet, however, provided no further benefit to those pigs that had received animal-based protein sources in the nursery (Fig. 3; Rodrigues et al., 2022b).

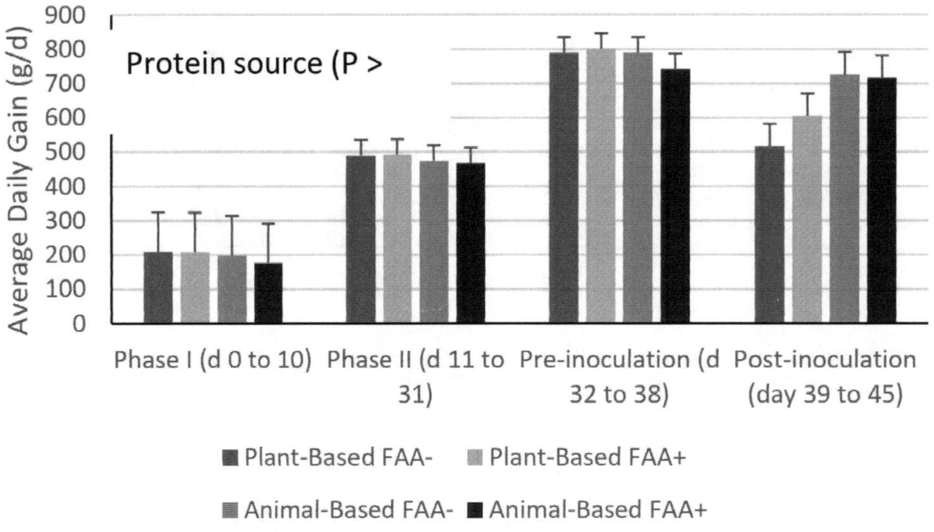

Figure 3. Growth performance of weaned pigs fed nursery diets containing either plant-based or animal-based protein sources and with (FAA+) or without (FAA-) a functional amino acid blend supplying methionine, threonine, and tryptophan at 120% of NRC (2012) requirements for growth. On d 32, piglets were placed on a common grower diet and inoculated with *Salmonella* Typhimurium on d 39 (n=8 pigs/treatment) (Rodrigues et al., 2022b).

The appropriate intervention may be dependent on the type of challenge experienced, and is likely related to the specific immune response to different challenge types and the proportion of the decrease in performance that is due to either reduced feed intake or alterations in nutrient utilization (Rodrigues et al., 2021a, Pastorelli et al., 2012). For example, while nutrient adjustment was effective in PRRS infected pigs, this adjustment was not effective in pigs challenged with Mycoplasma hyopneumoniae (Jasper, 2020). This may also explain why the blend of functional amino acids was effective during an enteric pathogen challenge (Rodrigues et al., 2021bc) but had reduced effectiveness during a sanitary challenge (van der Meer et al., 2016).

Conclusions

Disease challenge can have a substantial negative effect on productivity and profitability of swine production, however, there are a number of dietary strategies that can be used to improve pig robustness. Functional amino acids have been investigated for their role during disease challenge and ability to support both the immune response and growth performance. While these have been shown to be effective under some circumstances (i.e., normal birth weight, enteric challenge), this is not always the case. Overall, adjustments to dietary amino acid content is another tool that can be incorporated into pig health programs. Dietary strategies to improve pig health should, however, be based on the specific circumstances and goals of the production unit.

References

De Ridder, K., Levesque, C.L., Htoo, J.K., and de Lange, C.F.M. (2012) Immune system stimulation reduces the efficiency of tryptophan utilization for body protein deposition in growing pigs. J. Anim. Sci. 90:3485-3491.

Gabler, N.K. (2021) Feeding pigs during times of stress and disease. Proceedings of the International Conference on Swine Nutrition, Ames, IA, USA.

Valini, G.A., Righetti Arnaut, P., Franca, I., Trevisan Ortiz, M., Karpeggiane de Oliveira, M.J., Brandao Melo, A.D., Alves Marcal, D., Reis Furtado Campos, P.H., Htoo, J.K., Gastmann Brand, H., Hauschild, L. (2023) Increased dietary Trp, Thr, and Met supplementation improves growth performance and protein deposition of salmonella-challenged growing pigs under poor housing conditions. J. Anim. Sci. 101:skad141.

Jasper, J.E., Mendoza, O.F., Shull, C.M., Schweer, W.P., Schwartz, K.J., and Gabler, N.K. (2020) Increasing the ratio of SID lysine to metabolizable energy improves pig performance during a viral challenge. J. Anim. Sci. 98:skaa082.

Jasper, J.E. (2020) Evaluation of dietary approaches to improve growth performance of health-challenged pigs. MSc Thesis. Iowa State University, Ames, IA, USA.

Jayaraman, B., Htoo, J., and Nyachoti, C.M. (2015) Effects of dietary threonine:lysine ratios and sanitary conditions on performance, plasma urea nitrogen, plasam-free threonine and lysine of weaned pigs. Anim. Nutr. 1:283-288.

Koo, B., Choi, J., Yang, C., and Nyachoti, C.M. (2021) Diet complexity and L-threonine supplementation: effects on growth performance, immune response, intestinal barrier function, and microbial metabolites in nursery pigs. J. Anim. Sci. 98:skaa125.

Le Floc'h, N., LeBellego, L., Matte, J.J., Melchior, D., and Seve, B. (2009) The effect of sanitary status degradation and dietary tryptophan content on growth rate and tryptophan metabolism in weaning pigs. J. Anim. Sci. 87:1686-1694.

Litvak, N., Rakhshandeh, A., Htoo, J.K., and de Lange, C.F.M. (2023) Immune system stimulation increases the optimal dietary methionine to methionine plus cysteine ratio in growing pigs. J. Anim. Sci. 91:4188-4196.

McGilvray, W.D., Wooten, H., Rakhshandeh, A.R., Petry A., and Rakhshandeh A. (2019) Immune system stimulation increase dietary threonine requirements for protein deposition in growing pigs. J. Anim. Sci. 97:735-744.

Pastorelli H, van Milgen J, Lovatto P, and Montagne L (2012) Meta-analysis of feed intake and growth responses of growing pigs after a sanitary challenge. Animal. 6:952-961.

Rakhshandeh, A., Htoo, J.K. and de Lange, C.F.M. (2010) Immune system stimulation of growing pigs does not alter apparent ileal amino acid digestibility but reduces the ratio between whole body nitrogen and sulfur retention. Livest. Sci. 134:21-23.

Ramirez-Camba C.D., and Levesque C.L. (2023) The linear-logistic model: a novel paradigm for estimating dietary amino acid requirements. Animals. 13:1708.

Reeds, P.J., and Jahoor, F. (2001) The amino acid requirements of disease. Clin. Nutr. Suppl. 1:15-22

Reeds, P.J., Fjeld, C.R., and Jahoor, F. (1994) Do the differences between the amino acid compositions of acute-phase and muscle proteins have a bearing on nitrogen loss in traumatic states? J. Nutr. 124:906-910.

Rodrigues, L.A., Ferreira, F.N.A., Costa, M.O., Wellington, M.O., and Columbus, D.A. (2021a) Factors affecting performance response in pigs exposed to different challenge models: A multivariate approach. J. Anim. Sci. 99:skab035.

Rodrigues, L.A., Wellington, M.O., Gonzalez-Vega, J.C., Htoo, J.K., Van Kessel, A.G., and Columbus, D.A. (2021b) Functional amino acid supplementation, regardless of dietary protein content, improves growth performance and immune status of weaned pigs challenged with *Salmonella* Typhimurium. J. Anim. Sci. 99:skaa365.

Rodrigues, L.A., Wellington, M.O., Gonzalez-Vega, J.C., Htoo, J.K., Van Kessel, A.G., and Columbus, D.A. (2021c) A longer adaptation period to a functional amino acid-supplemented diet improves growth performance and immune status of *Salmonella* Typhimurium-challenged pigs. J. Anim. Sci. 99:skab146.

Rodrigues, L.A., Panisson, J.C., Kpogo, L.A., Gonzalez-Vega, J.C., Htoo, J.K., Van Kessel, A.G., and Columbus, D.A. (2022a) Functional amino acid supplementation post-weaning mitigates the response of normal birth weight more than for low birth weight pigs to a subsequent *Salmonella* challenge. Animal. 16:100566.

Rodrigues, L.A., Panisson, J.C., Van Kessel, A.G., and Columbus, D.A. (2022b) Functional amino acid supplementation attenuates the negative effects of plant-based nursery diets on the response of pigs to a subsequent *Salmonella* Typhimurium challenge. J. Anim. Sci. 100:skac267.

Schweer, W.P., Mendoza, O.F., Shull, C.M., Lehman, J., Gaines, A.M., Schwartz, K.J., and Gabler, N.K. (2019) Increased lysine:metabolizable energy ratio improves grower pig performance during a porcine reproductive and respiratory syndrome virus challenge. Trans. Anim. Sci. 3:393-407.

Trevisi, P., Corrent E., Mazzoni, M., Messori, S., Priori, D., Gherpelli, Y., Simongiovanni, A., and Bosi, P. (2015) Effect of added dietary threonine on growth performance, health, immunity and gastrointestinal function of weaning pigs with differing genetic susceptibility to E. coli K88ac. J. Anim. Physiol. Anim. N. 511-520.

Wellington, M.O., Thiessen, R.B., Van Kessel, A.G., and Columbus, D.A. (2020a) Intestinal health and threonine requirement of growing pigs fed diets containing high dietary fibre and fermentable protein. Animals. 10(11), 2055.

Wellington, M.O., Hamonic, K., Krone, J.E.C., Htoo, J.K., Van Kessel, A.G., and Columbus, D.A. (2020b) Effect of dietary fiber and threonine content on intestinal barrier function in pigs challenged with either systemic *E. coli* lipopolysaccharide or enteric *Salmonella typhimurium*. J. Anim. Sci. Biotechnol. 11:38.

Wellington, M.O., Agyekum, A.K., Hamonic, K., Htoo, J.K., Van Kessel, A.G., and Columbus, D.A. (2019) Effect of supplemental threonine above requirement on growth performance of *Salmonella typhimurium* challenged pigs fed high fiber diets. J. Anim. Sci. 97:3636-3647.

Wellington, M.O., Htoo, J.K., Van Kessel, A.G., and Columbus, D.A. (2018) Impact of high dietary fiber and immune system stimulation on threonine requirement for whole body protein deposition in growing pigs. J. Anim. Sci. 96:5222-5232.

Valini, G.A.C., Arnaut, P.R., Franca, I., Ortiz, M.T., Oliveira, M.J.K., Melo, A.D.B., Marcal, D.A., Campos, PH.R.F., Htoo, J.K., Brand, H.G., and Hauschild, L. (2023) Increased dietary Trp, Thr, and Met supplementation improves growth performance and protein deposition of salmonella-challenged growing pigs under poor housing conditions. J. Anim. Sci. 101:skad141.

van der Meer, Y., Lammers, A., Janasman, A.J.M., Rijnen, M.M.J.A., Hendriks, W.H., and Gerrits, W.J.J. (2016) Performance of pigs kept under different sanitary conditions affected by protein intake and amino acid supplementation. J. Anim. Sci. 94:4704-4719.

Nutrition & Management of Weaned Pigs Fed Low Zinc Diets

Mike Tokach, Jordan Gebhardt, Jason Woodworth, Joel DeRouchey, and Bob Goodband

Applied Swine Nutrition Team, Kansas State University, 246 Weber Hall, Manhattan, KS 66506, USA, 785-532-2032, mtokach@ksu.edu

1. Introduction

Since the first publication demonstrating the value of ZnO in reducing post-weaning diarrhea (Paulsen), zinc oxide use has become a core component of most nursery diet programs. When included in the diet at 1,500 to 3,000 ppm for the first 2 to 3 weeks after weaning, Zn from ZnO improves feed intake and growth performance. Zinc has numerous modes of action that combined result in reduction in diarrhea and improvement in feed intake and growth rate after weaning. The modes of action can be grouped into four categories: high zinc requirement; gut protection; antibacterial; and direct impacts on feed intake. When replacing zinc in the diet, the main modes of action for ZnO should all be considered.

2. Management Needs For Pigs Fed Low Zinc Diets

Several management changes that should be considered are outlined in Table 1. Increasing weaning age to have a more mature digestive system must be considered. The reduction in mortality that comes with increasing weaning age makes this an important consideration. Adam Moeser's work demonstrating that the mode of action is largely gut mediated suggests that increasing weaning age should greatly reduce the need for ZnO use. Pigs that are older at weaning (ex. 24 to 26 days) are able to handle an E coli challenge with less diarrhea and translocation of the toxins across the intestine into the bloodstream. These findings are confirmed by producers around the world that have found that increasing weaning age has been required to reduce mortality when ZnO can no longer be fed at pharmacological levels.

Table 1: Nutrition %management strategies to consider for weaned pigs fed low zinc diets

Management
- Increase weaning age
- Ensure adequate sanitation
- Reduce management stressors (temperature, humidity, fill time, number of pig sources, etc)
- Enhance immunity (colostrum intake and vaccines)
Genetics
- Use of genetics with *E. coli* resistance or more robust GI tract
Diets
- Adopt and feed low acid binding capacity (ABC-4) diets
o Lowers buffering capacity of diet.
- Reduce dietary crude protein as much as possible with use of feed grade amino acids maintaining a maximum digestible lysine to crude protein ratio of 6.35%.
o Dietary lysine will likely need to be below the requirement of most modern genotypes.
- Increase amino acid ratios relative to lysine when diets are below the lysine requirement. For example: 60% Met&Cys:Lys; 21% Trp:Lys; 65% Thr:Lys; 72% Val:Lys; 55% Ile:Lys.
- Use highly digestible protein sources immediately after weaning.
- Include high levels of dietary phytase to improve protein digestibility and lower calcium content
- Minimize soluble nonstarch polysaccharide levels
- Include a source of insoluble nonstarch polysaccharides (ex. lignocellulose, oat hulls, coarse wheat bran)
- Consider grinding grains coarser
- Consider inclusion of xylanase
- Review vitamin, antioxidants, and fatty acid levels required for optimal immune function
- Review inclusion of other feed additives if lower levels of ZnO must be fed
o Add copper to near the legal limit
o Consider another zinc source that may have improved efficacy relative to ZnO when added at lower levels (ex. Zn nanoparticles)
o Inclusion of acidifiers, short or medium chain fatty acids, or other antibacterial additives
Water quality and pH
- Lower water pH if over 7.0
- High mineral water will increase buffering capacity and increase loose stools

Sanitation to minimize bacteria remaining in the environment is always important; however, the importance becomes enhanced when ZnO is not available to help control the pig's response to the bacteria. Decreasing the bacteria remaining on feeders, drinkers, flooring, and mats is critical to minimize the amount of bacteria that the pig is exposed to in the nursery.

Because stressors are additive in their negative impacts, reduction in other management stressors also provides benefit. Steps such as provision of creep feed or socialization of pigs prior to weaning can lessen the stress that the pig experiences at weaning. Minimizing fill time also improves performance and reduces the number of pigs requiring antibiotic treatments, even when sourced from the same sow farm. Socializing pigs prior to weaning also helps with maintenance of intake as pigs transition to the nursery. Proper temperature, ventilation, and immunization are also critical steps in lessening other stressors.

Because some pigs are resistant to *E coli* or nursery feed intake patterns that make them less susceptible to problems with *E coli*, genetic changes may be necessary in some production systems if they cannot control postweaning diarrhea by other means. Different genetic lines also have different intake patterns after weaning. Pigs that have very low intake immediately after weaning can have more gut health challenge because the low intake allows pathogens to colonize in the gut.

3. Nutrition Changes for Low Zinc Diets

Numerous dietary changes have been shown to be helpful to reduce postweaning diarrhea. A review by Pluske et al provides an excellent summary of dietary changes that should be considered for optimal gut health. Many of their recommendations are included in Table 1. The dietary changes revolve around minimizing the amount of undigested material available for bacterial growth, increasing passage rate and lowering gastric pH to limit bacterial growth, improving immune function, and providing other additives that may have antibacterial effects. Lowering dietary crude protein and the acid binding capacity of the diet are key components to increase fecal dry matter and reduce diarrhea when ZnO is not fed at pharmacological levels. Our research has demonstrated that lowering ABC-4 (amount of acid required to lower dietary pH to 4.0) to approximately 200 meq/kg increases fecal dry matter and growth performance linearly when pharmacological levels of Zn are not included in the diet immediately after weaning. Although the acid binding capacity concept has been around for years, it was not considered important when high levels of ZnO could be fed; however, recent publications on the background and ABC-4 loading values for different feed ingredient has brought this concept back to the forefront (Stas et al. 2022; XXX et al., 2022).

Inclusion of other metals or alternative zinc sources also should be considered. Pharmacological levels of copper and zinc are usually not additive in their response; however, if high levels of zinc are not included in the diet and pharmacological levels of copper remain legal, copper can provide benefit at growth promoting levels. Similarly, if ZnO cannot be fed at high levels, other zinc sources may provide more benefit than ZnO when the inclusion rate is limited to 500 ppm or less. Part of antibacterial effect of ZnO is thought to be due to its ability to generate hydrogen peroxide from its surface[33]. Similar to the benefit of reducing feed particle size increasing surface area and opportunities for enzymes to improve digestibility, increases in Zn surface area through lower particle size increases the efficiency of hydrogen peroxide production and antibacterial activity. Some researchers found that lower inclusion rate of porous and nanoparticles of ZnO could achieve similar performance to higher inclusion rate of normal ZnO.

Water quality and pH also should be evaluated. If pH is over 7.0, reducing pH with water acidification is often necessary to control postweaning diarrhea. Field reports from Denmark suggest lowering water pH to 5.0 may be beneficial when hemolytic *E coli* is a problem. If the water is high in sulfates or total mineral content, it can result in loose watery feces and the visual appearance of diarrhea. High sulfate water does not appear to have a major impact on pig growth; however, it may increase pig treatments and frustration in the barn due to inability to control the loose stools that occur. In these situations, obtaining an alternative water source may be the preferred solution. Lowering water pH to 5.0 or less appears to be important in controlling diarrhea when high levels of ZnO can't be used.

4. Conclusion

Research and field experience from around the world has demonstrated that a combination of management and dietary changes are required to minimize diarrhea when pharmacological levels of zinc cannot be fed.

References

Poulsen, H.H.D. (2018) Zinc oxide for weaned pigs. In: Proceedings of the 40th annual meeting of the European Association for Animal Production. EAAP Publications, Dublin. 265-266.

Pluske, J.R. Turpin, D.L., and Kim, J.C. (2018) Gastrointestinal tract (gut) health in the young pig. Animal Nutrition. 4:187-196. https://doi.org/10.1016/j.aninu.2017.12.004

Stas, E.B. Tokach, M.D. DeRouchey, J.M. Goodband, R.D., Woodworth, J.C., and Gebhardt, J.T. (2023) Evaluation of the acid-binding capacity of ingredients and complete diets commonly used for weanling pigs. Translational Animal Science, 2022, **6, 1–9. https://doi.org/10.1093/tas/txac104**

Wang, L.F., Bergstrom, J.R., Hahn, J.D., Young, M.G., and Zijlstra, R.T. (2023) Acid-binding capacity of feed in swine nutrition. Animal Feed Science and Technology. 295:115519. https://doi.org/10.1016/j.anifeedsci.2022.115519

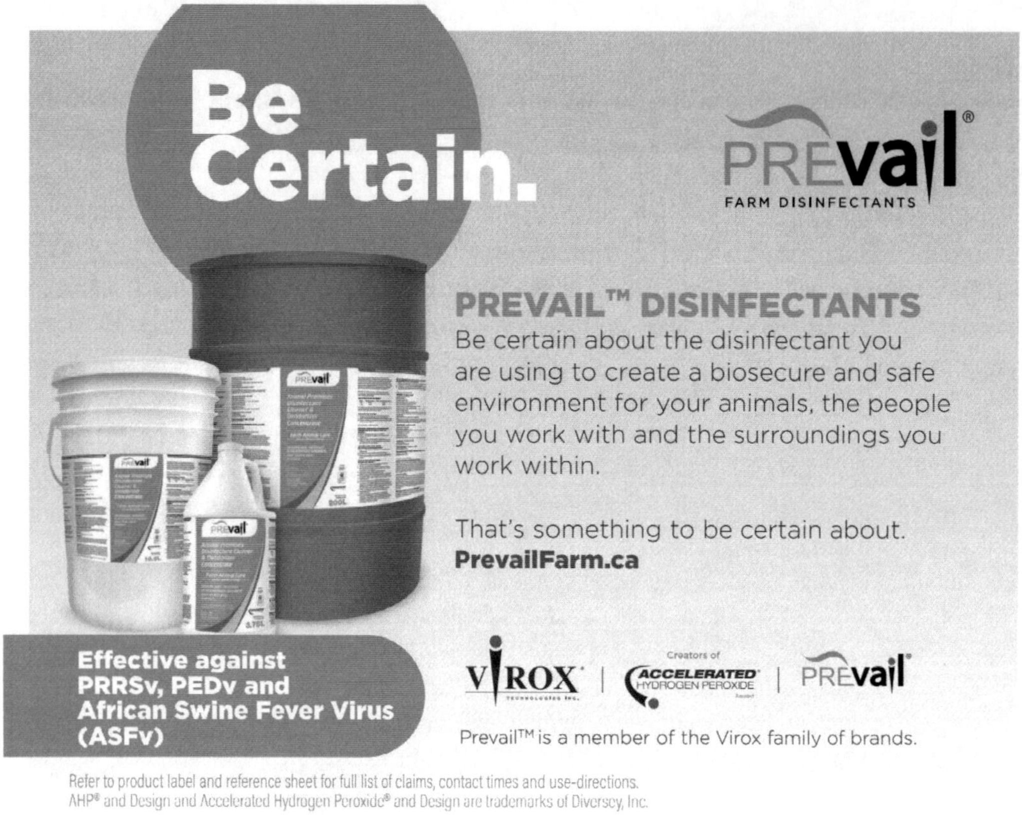

Be Certain.
PREvail®
FARM DISINFECTANTS
PREVAIL™ DISINFECTANTS
Be certain about the disinfectant you are using to create a biosecure and safe environment for your animals, the people you work with and the surroundings you work within.
That's something to be certain about.
PrevailFarm.ca
Effective against PRRSv, PEDv and African Swine Fever Virus (ASFv)
VIROX
Creators of
ACCELERATED HYDROGEN PEROXIDE
PREvail
Prevail™ is a member of the Virox family of brands.
Refer to product label and reference sheet for full list of claims, contact times and use-directions.
AHP® and Design and Accelerated Hydrogen Peroxide® and Design are trademarks of Diversey, Inc.

GSF
SOW FEEDING
FEED SOWS IN PEN GESTATION INDIVIDUALLY.
• Customize feed requirements of each individual sow.
• Extensive feed management control.
• Uses RFID technology to identify sows.
• Adjust each sows ration for optimal nutrition.
• Flexible modular design.
AP
UNLOCK BETTER PERFORMANCE
AUTOMATEDPRODUCTION.COM

Weaning Age Impact on Piglet and Sow Performance

Luc Dufresne DVM

Introduction: A Trip To Memory Lane

What is the optimum age to wean pigs? That question has been on the lips of swine producers since the start of modern intensive swine production. Naturally, weaning age is a process that happens gradually around 10 to 12 weeks of age. Considering the farrowing crate was the most expensive asset in a sow farm, producers tried to reduce wean age gradually from 6 weeks to 4 weeks, which was the acceptable practice in intensive swine production until 1990. Then, two important events happened:

1. The development of medicated early weaning by Dr. Tom Alexander (Alexander et al., 1980) allowed the elimination of several diseases on piglets from infected dams by treating them with antibiotics and weaning them at a very young age (7-10 days)
2. Significant advancements in swine nutrition allowed the development of a more complex diet with more digestible ingredients like spray-dried plasma that was more digestible for the young piglet.

Those two events propelled the rise of Segregated Early Weaning (SEW) as one of the leading production strategies of the 21st century (Harris, 2000). This also allowed the promotion of the development of larger sow farms and the development of the more extensive production system sold on the assumptions that:

1. A large early wean sow farm will produce a large amount of high-health pigs
2. Litter size reduction associated with earlier wean age will be offset by an increase in the number of litter farrow in a year.
3. Increase utilization of the most expensive asset in the sow farm: the farrowing crate.
4. If you have the correct diet and the correct environment, the weaning age has a minimum effect on performance.
5. The higher health status of the younger weaned pig will outweigh the negative impact of early weaning.
6. SEW was the answer to long-term health and production stability!

SEW felt short of its initial promises! In its defense, the principle was bastardized, stretched to the utmost limit, and broke. While we can see some of the pitfalls and shortcuts that were taken and caused the failure on the health side, we believe at the time that our understanding of the weaning age and its impact was well understood and accounted for.

Main et al. (2004) showed that we were wrong again and that wean age was correlated with lifelong ADG and mortality. This work was pivotal in how producers and swine experts started revisiting the weaning age and its impact on pig health and productivity. Further research shows that it was primarily driven by the impact of early weaning on GI tract long-term functionality.

The GI Tract

The GI tract mucosa plays several critical roles during the early life of the piglets.

- Absorption of nutrients
- Providing an effective barrier to pathogens, toxins, and antigenic components in the gut.
- Balance immune development on food and microbial antigens

The GI system has multiple layers of sophisticated barrier mechanisms to perform these divergent functions.

- The GI Epithelium:

The GI Epithelium is a single layer of epithelial cells lining the GI tract and represents the most extensive contact between the host and the outside world. The intestinal epithelium facilitates the breakdown and uptake of nutrients while facilitating a massive amount of water movement. At the same time, the epithelial must provide a barrier to pathogens and antigens toxins and serve as the first line of defense. One of the most critical mechanisms of the defense supplied by the GI epithelium is the permeability barrier, mainly controlled by the tight junction, which consists of numerous intracellular intercellular membrane proteins.

- The GI immune system

The GI immune system is the largest immune organ in the body. This system is highly regulated and maintains a delicate balance between controlling the inflammatory response to antigenic substances and reactiveness to infection by a pathogen or breach in the epithelial barrier. This is critical for good GI health, and imbalance plays a significant role in GI inflammatory disorder.

Several maternal and host mechanisms inhibit immune activation during early GI development, indicating the importance of an immunosuppressive state for the good development of the immune system. Therefore, excessive immune stimulation during this critical period could negatively affect the development and function of the gut immune system.

- Enteric nervous system

The enteric nervous plays a central role in gut and overall systemic health. The enteric nervous system consists of two major neural ganglia in the muscle and submucosa, which control motility and peristalsis. The enteric nervous system comes up through the constant release of an area of neurochemicals that play a central role in gut motility—suppression in absorption and modulation of epithelial permeability. Toxins can be sensed by sensory fiber under the epithelium and initiate neural secretory reflexes via the release of vasoactive intestinal peptides that drive electrolyte and fluid secretion. This Neuroimmune communication also plays a role in GI diseases like IBS, where an increase in nerve mast-cell interaction is detected (Barbara et al. 2004,2007). Interestingly, a similar rise in nerve mast-cell interaction is observed in adult pigs that were early wean (16 days) vs. late wean pigs (28 days) (Pohl et al., 2017).

Impact of Weaning Age on GI Tract Development & Function

Weaning is a gradual process that would approach completion around 10 to 12 weeks of age and would coincide with the maturation of the GI epithelial immune and nervous system. However, weaning in commercial pig production is one of the most stressful events in a pig's life. Weaning is abrupt, occurring between 14 to 28 days of age. Not only is it a sudden separation from the mother and a drastic dietary change, but it is also compounded by other stress: transport, mixing/fighting, vaccination, etc. Commercial weaning also coincides with declining passive immunity. All those stresses associated with early weaning have the potential to disrupt the normal epithelial immune and enteric nervous system development processes:

- Breakdown of the GI barrier function
    - Epithelial barrier breakdown:
        - Moser et al. 2007 demonstrated that compared with age-match litter maid pigs, early-weaned pigs show an increase in intestinal permeability
        - This change in permeability has been shown to persist up to 9 weeks post-weaning (Smith et al., 2010) and even into adulthood (Pohl et al., 2017)

    - GI immune barrier breakdown
        - following weaning, an upregulation of the proinflammatory cytokines indicates an activation of the GI immune system (Hu et al., 2013)

- Activation of mast cells that plays a critical role in intestinal barrier breakdown observed in early weaned pigs (Moser et al., 2007)

- GI immune function
  - Early weaning also appears to create long-lasting functional changes in the GI immune function and disease susceptibility.
  - Early-weaning piglets have increased susceptibility to enterotoxigenic E. coli challenge (Mclamb et al., 2013)
  - Early wean pigs show a lower immune response to stress later in life (Davis et al., 2006)
  - Early weaning creates a compromised immune response characterized in the gut by mast cell Hyperplasia in the small and large intestines 2- and 9-week post-weaning.

    (Mclamb et al. 2013) and still present in Adults (Pohl et al., 2017)
  - As they age, late-weaning pigs see an expected decline in enteric neuron numbers, while early-weaning pigs do not show that reduction. (Medland et al 2016). This demonstrated that early weaning induced persistent upregulation of the enteric cholinergic system, showing that early wean pigs will be more susceptible to hypersecretion, explaining the increased vulnerability of early wean piglets to enterotoxigenic E. coli.

- Mast Cells
    - Mast Cells are required for host defense and wound healing, but excessive activation of mast cells is a central mechanism in allergy, asthma, and IBS in humans (Hamilton et al., 2016)
    - Early wean pigs show a higher mast cell activation and develop a persistent intestinal mast cell hyperplasia that is still evident nine weeks post-weaning (Moser et al., 2007; Smith et al., 2010)
    - This is a significant factor in increasing gut permeability (Moser et al., 2007) and plays a central role in increased epithelial permeability disturbance associated with early weaning.

- Take-home lesson:
    - The GI tract is critical for nutrient absorption and immune regulation.
    - Commercial weaning is a very stressful event that has a long-lasting effect on the GI immune barrier. Creating long-lasting issues locally in the GI tract but also systemically due to the resulting immune system impairment
    - This effect is more pronounced and longer lasting the younger the pig is weaned.
    - Effects are less significant and more transient when pigs are weaned at 28 days or beyond.

Wean Age Impact on Post-Weaning Performance: Where The Tire Meets The Road

Main et al. (2004) did the first control study looking at age at weaning on wean-to-finish performance. Main showed linear improvement of both gain and livability in two different trials as wean age increased. Similar findings were observed in three subsequent control Studies (Smith et al., 2008; Faccin et al., 2020; Faccin et al., 2020b). These findings were corroborated by extensive data group analysis (Rosero et al., 2016); those findings give some robust evidence of the impact of wean age on grow-finish performance. The table below is the combination of the four control studies' linear improvement in performance for each day increase in weaning age. For those four studies, the linear improvement is overall pretty similar for weaning weight (220 to 300 gr per day increase),42 days post-weaning (.62 to .93 kg per day increase), and wean to finish ADG (10 to 13 gram per day increase). Wean-to-finish mortality is the only parameter showing significant variation, where two studies show significant improvement and two show no to limited progress with increased wean age. This is not unexpected as it is assumed that livability improvement would have more impact in more health-challenging settings. This hypothesis is supported by a large data study (Rosero et al., 2016)

Table 1: Linear improvement associated with an increase in weaning age

study	Main et al. 2004	Smith et al. 2008	Faccin et al. 2020	Faccin et al. 2020 b
wean age range (days)	12 to 21	15 to 20	18.5 to 24.5	19 t0 28
wean weight gr	257	300	220	250
D 42 post-weaning weight KG	0.93	0.62	0.928	0.802
Wean to finish ADG gr/day	9.9		11.1	13
Wean to finish Mortality	0.47%	0.10%		0.77%
WT sold oer pig Wean Kg	1.8		0.71	2.21

Lactation Length Impact on the Sow Performance

Several retrospective data studies have shown the impact of weaning on service intervale and subsequent litter size (AuMaitre et al., 1976; LeCozier et al., 1997; Dewey et al .1994; Costa et al., 2004). Most of those studies were based on analyzing large databanks and looking at predictors of subsequent litter size. In most, the wean-to-service interval was a good predictor of litter size; sows bred five days or less after weaning had the highest total born. The other factor associated with litter size was lactation length, which is logical since sows with longer lactation lengths generally have shorter wean-to-service intervals. The general linear assumption is that each day's increase in lactation length will give you an extra .1 piglet total born. Note that some control studies looking at lactation length's impact on subsequent litter size have shown minimum to no effect of lactation length on following litter size (Smith et al., 2008). Nevertheless, 0.1 pig per day of lactation remained the acceptable assumption.

Putting It All Together (The Economic Model)

We are circling back to the initial question: What is the optimum wean age? More specifically, what is the optimum wean age for my farm or my operation? The technique that is often used in that situation is partial budgeting. Partial budgeting is a planning and decision-making framework used to compare the costs and benefits of alternatives faced by a farm business. It focuses only on the changes in income and expenses resulting from implementing a specific alternative. Thus, all aspects of farm profits that are unchanged by the decision can be safely ignored. It allows you to understand better how an option will affect the farm's profitability. Like in most models, the value of partial budget analysis is dependent upon the quality of the information used for the analysis; in other words, garbage is in, garbage is out.

The tables below show the primary assumption used for the model.

Table 2:Linear change per day of age/lactation

Parameter	Impact per days
Wean to service interval	.167 days
Conception rate	.167%
Fall out rate	.083%
Total born	0.1 piglet
Stills & mummies	.033%
Preweaning mortality	.167%
Lactation feed intake	.1 kg
Weaning weight	250 gr
Quantity of Prestarter Kg	-250 gr
Weight at the exit of the nursery (42 days)	800 gr
Wean to finish the mortality	.33%
Weight sold at market	1.75 kg

Table 3: Inventory value

Piglet Birth	$	42.30
Piglet Processing	$	44.65
Piglet Weaning	$	47.00
Nursery Early	$	51.00
Nursery Late	$	60.00
Finisher Early	$	83.00
Finisher Mid	$	125.00
Finisher Late	$	185.00
Gilt	$	370.00
Sow	$	420.00

Table 4: Feed cost

Gestation	**385.0 $**
Lactation	**475.0 $**
Prestarter	**1,100.0 $**
Nursery1	**800.0 $**
Nursery2	**515.0 $**
Nursery3	**475.0 $**
Finisher 1	**440.0 $**
Finisher 2	**430.0 $**
Finisher 3	**400.0 $**
Finisher 4	**400.0 $**

Table 5: Other costs

Sow farm	
New Construction	**$4,000.00**
Labor cost per sow per year	**$ 170.00**
Breeding and genetic cost	**$ 120.00**
nursery place	
New Construction	**$375.00**
contract nursery cost day	**$ 0.12**
Finisher Space	
New Construction	**$650.00**
Contract finisher cost per day	**$0.16**
Carcass market price per KG	**$ 1.90**

In this partial budgeting model, we will look at two options for 3200 farrow to finish the system currently weaning at an average age of 18 days:

- Option A: Increase weaning age from 18 to 21 days by increasing farrow crate efficiency (reducing downtime by one day and increasing transfer in the farrowing room from 110 to 112 days of gestation
- Option B: Increasing the weaning age from 18 to 24 days by doing the same as above and adding 60 extra farrow crates. The 60 crates addition will cost $378,000 with financing and will be amortized over seven years, making it an extra cost of $17.00 per sow per year.
- The two options show a similar pig per sow per year as the lower litter per year is somewhat offset by the higher total born with higher lactation length.
- In that simulation, we also add some extra space available in Breeding gestation, allowing us to increase the herd inventory to maintain the weekly farrowing target
- Maintaining PSY and increasing the sow inventory slightly allow the farm total throughput to increase in each option.
- The increased revenue is offsite by the increased feed cost mainly associated with the increased lactation length and, to a lesser extent, the inventory increase.
- We also incurred an added cost on breeding herd and labor for the higher inventory.
- Overall, option A increases the wean pig cost by 0.29$ while option B increases it by $1.35, mainly due to the cost of the added farrowing crates

	control	Option A wean age 21 increase farrowing crate efficiency	difference	Option B farrowing crate efficiency add 60 extra farrowing	difference
		barn capacity			
total gestation crate space	1300	1300	0	1300	0
total gestation pen space	1600	1600	0	1600	0
total farrowing crate	540	540	0	600	60
total space	3440	3440	0	3500	60
space efficiency in gestation	92%	95%	3%	95%	3%
intervention cost per sow per year	$ -		$ -	$ 17.00	$ 17.00
wean age	18	21	3	24	6
open crate days	8	5	-3	5	-3
wean to service interval	6	5.5	-1	5.2	-1
conception rate	91.5%	92.0%	0.5%	92.5%	1.0%
post pregcheck fall out	4.00%	3.75%	-0.25%	3.50%	-0.50%
Farrowing rate	87.5%	88.3%	1%	89%	2%
Litter per sow per year	2.41	2.37	-0.03	2.34	-0.07
total born	15	15.3	0.3	15.6	0.60
Stills & mummies	8.0%	8.1%	0%	8.2%	0%
Born alive	13.8	14.06	0.26	14.33	0.53
preweaning mortality	12.00%	12.50%	0.50%	13.00%	1.00%
wean per litter	12.14	12.30	0.16	12.47	0.32
PSY	29.21	29.19	-0.03	29.13	-0.08
average daily lactation itake Kg	6.00	6.30	0.30	6.60	0.60
feed cost per sow per year	$ 464.01	$ 482.78	$ 18.77	$ 502.88	$ 38.88
inventory	3143	3187	44	3223	80
PIG WEAN PER YEAR	91,809	93,012	1,203	93,881	2072
Wean pig revenue	$ 55.00	$ 55.00	$ 66,159	$ 55.00	$ 113,979
feed cost difference	$ 1,458,222.98	$ 1,538,560.69	$ (80,338)	$ 1,620,729.46	$(162,506)
inventory change on breeding herd cost	$ 377,121.80	$ 382,428.30	$ (5,306)	$ 10,716,043.08	$ (9,623)
inventory change impact on labor	$ 534,255.88	$ 541,773.42	$ (7,518)	$ 547,887.92	$ (13,632)
Treatment cost	0	$	-	$	(54,788.79)
difference in total profit lost			$ (27,003)		$(126,571)
per pig			$ (0.29)		$ (1.35)
per sow			$ (8.47)		$ (39.27)

- We then continue to run those two options into the growing phase using two models, one looking at a fixed time with variable marketing weight and the other with a fixed weight with variable growing time.
- For both options, we assume 75% of the mortality gain is recuperated in the nursery phase while 25% is captured in the finisher phase
- The increased weaning weight also allows a reduction of pre-starter usage and a significant cost reduction during the nursery phase

- The combination of increased sow inventory and reduced mortality allows the farm to market 2000 more pigs in option A and 3800 pigs in option B.
- In the fixed time options, the overall linear improvement in gain increases the market live weight by 5.25 and 10.5 kg, respectively. The increase in weight and pig marketed are driving the increase in revenue by nearly $400,000 ($4.14 per pig) for option A and 728,000 ($7.68) for option B.
- In the fixed weight options, the overall financial benefit is less since higher revenue due to higher weight is not achieved. However, the increase is still significant at roughly $250,000 ($2.68) and $450,000 ($4.72).
- In both options, in the economic parameters of the model and assuming the linear increase of performance is achieved, increasing the weaning age creates a significant revenue increase that more than offsets the increase in wean pig cost.

Nursery					
	18 day wean age	increase FC efficiency increase wean age to 21 days	diff	add 60 FC increase wean age 24 days	diff
pig transfer	91,809	93012	1,203	93991	2,182
pig transfer cost	$ 45.00	$ 45.29	$ 0.29	$ 46.35	$ 1.35
nursery stage	42	42	0	42	0
entry weight (KG)	5.250	6.000	0.750	6.750	1.500
nursery ADG (KG)	0.400	0.439	0.039	0.479	0.079
nursery FE	1.40	1.42	0.02	1.44	0.04
transfer weight (KG)	22.05	24.45	2.40	26.85	4.80
prestarter (KG)	1.75	1.25	-0.50	0.50	-1.25
starter 1 (KG)	5	5	0	5	0
starter 2 (KG)	10	10	0	10	0
starter 3 (KG)	6.77	9.949	3	13.444	7
feed cost	$ 14.29	$ 15.25	$ 0.96	$ 16.09	$ 1.80
grow cost	$ 6.04	$ 6.04	$ -	$ 6.04	$ -
mortality	4.00%	3.25%	-0.75%	2.50%	-1.50%
transfer out	88,137	89,989	1,852	91,641	3,505
wean pig cost	$ 4,131,405	$ 4,212,513	$ 81,108	$ 4,356,483	$ 225,078
total feed cost	$ 1,259,539	$ 1,372,404	$ 112,865	$ 1,474,132	$ 214,593
total grow cost	$ 554,627	$ 561,894	$ 7,267	$ 567,809	$ 13,182

Finisher fix days					
	18 day wean age	increase FC efficiency increase wean age to 21 days	diff	add 60 FC increase wean age 24 days	diff
finisher stage days	105	105	0	105	0
Finisher ADG (KG)	0.870	0.897	0.027	0.924	0.054
finisher FE	2.6	2.62	0.02	2.64	0.04
market weight	113.4	118.65	5.25	123.9	10.5
market price carcass	$ 1.90	$ 1.90	0	$ 1.90	0
yield	80%	80%	0	80%	0
average index	110	110	0	110	0
revenue per pig	$ 189.60	$ 198.38	$ 8.78	$ 207.16	$ 17.56
finisher 1	30	30	0	30	0
finisher 2	50	50	0	50	0
finisher 3	75	75	0	75	0
finisher 4	82.5	91.8	9.3	101.2	18.7
feed cost	$ 97.70	$ 101.42	$ 3.72	$ 105.18	$ 7.48
grow cost	$ 17.26	$ 17.26	$ -	$ 17.26	$ -
mortality	4.00%	3.75%	-0.25%	3.50%	-0.50%
transfer out	84611	86615	2,003	88434	3,823
total feed cost	$ 8,266,850	$ 8,784,583	$ 517,733	$ 9,301,890	$ 1,035,040
total grow cost	$ 1,521,263	$ 1,553,237	31,974	$ 1,581,753	$ 60,490
total gross revenue	$ 16,042,685	$ 17,182,831	1,140,146	$ 18,320,013	$ 2,277,328
Total net revenue	$ 309,001	$ 698,199	389,198	$ 1,037,948	$ 728,946
revenue net per wean pigs	$ 3.37	$ 7.51	$ 4.14	$ 11.04	$ 7.68

Finisher fix weight					
	18 day wean age	increase FC efficiency increase wean age to 21 days	diff	add 60 FC increase wean age 24 days	diff
finisher stage days	105	99	-6	94	-11
Finisher ADG (KG)	0.870	0.897	0.027	0.924	0.054
finisher FE	2.6	2.6	0.00	2.6	0.00
market weight	113.4	113.4	0.00	113.4	0.0
market price carcass	$ 1.90	$ 1.90	0	1.9	0
yield	80%	80%	0	80%	0
average index	110	110	0	110	0
revenue per pig	$ 189.60	$ 189.60	$ -	$ 189.60	$ -
finisher 1	30	30	0	30	0
finisher 2	50	50	0	50	0
finisher 3	75	75	0	75	0
finisher 4	82.51	76.27	-6.2	70.03	-12.5
feed cost	$ 97.70	$ 95.21	$ (2.50)	$ 92.71	$ (4.99)
grow cost	$ 17.26	$ 16.30	$ (0.96)	$ 15.39	$ (1.87)
mortality	4.00%	3.75%	-0.25%	3.50%	-0.50%
transfer out	84611	86615	2,003	88434	3,823
total feed cost	$ 8,266,850.18	$ 8,246,395.07	$ (20,455)	$ 8,198,872.81	$ (67,977)
total grow cost	$ 1,521,263	$ 1,466,671	-54,592	$ 1,410,620	$ (110,642)
total gross revenue	$ 16,042,685	$ 16,422,528	379,844	$ 16,767,470	$ 724,785
Total net revenue	$ 309,001	$ 562,651	253,649	$ 759,553	$ 450,552
revenue per wean pig	$ 3.37	$ 6.05	$ 2.68	$ 8.08	$ 4.72

For this exercise. I use the assumption of weaning age between 18 and 24 because that is the age range in which I was the most comfortable with the linear improvement in performance. The control research data is primarily for pigs weaned between 12 and 25 days of age. As we go outside those parameters, assuming the performance rate would remain the same, especially if piglets are weaned above 28 days, is more complicated. As several published researches show, the stress of weaning is less impactful and more transient as weaning age increase one would assume that improvement should start tampering off as wean age increase but no data are currently available to determine when that inflexion point happen.

Conclusion

Weaning is one of the most stressful events in a pig's life. Although piglets overcome weaning and can perform well, new research has demonstrated that early life stressors such as weaning altered the development of the GI barrier function, leading to long-lasting negative consequences on gut health and disease susceptibility of the animal throughout its life. Research has shown that the impact of weaning age on zootechnical performance was seen throughout animal production life, and that impact was linear, becoming less significant as weaning age increased. The partial model of a 3200 sow farrow to finish system shows, within the parameter of the model and linear improvement assumption, that increasing the weaning age from 18 to 21 and 24 by improving crate efficiency and adding farrow crate capacity would increase both the farm total revenue and the revenue per pigs. More research is needed to determine the linear improvement and economic benefit of increasing the weaning age above 28 days.

References

Alexander TJ, Thorton K, Boon G, Lysons RJ, and Gush AF. Medicated Early Weaning to obtain pig free from pathogens endemic in the herd of origin. The Veterinary Record. 1980;106: 114-119

Barbara G., Stanghellini V., De Giorgio R., Cremon C., Cottrell G.S., Santini D. Activated mast cells near colonic nerves correlate with abdominal pain in irritable bowel syndrome. Gastroenterology. 2004; 126:693–702.

Barbara G., Wang B., Stanghellini V., de Giorgio R., Cremon C., Di Nardo G. Mast cell-dependent excitation of visceral-nociceptive sensory neurons in irritable bowel syndrome. Gastroenterology. 2007; 132:26–37.

Costa E.P, Filha A., Costa A.H.A., Carvalho F.F. Santos A.K. Silva A.F. Influence of the lactation length in the subsequent litter size of sows Anim Reprod, 2004; 1:.111-114,

Davis M.E., Sears S.C., Apple J.K., Maxwell C.V., Johnson Z.B. Effect of weaning age and commingling after the nursery phase of pigs in a wean-to-finish facility on growth, and humoral and behavioral well-being indicators. J Anim Sci. 2006; 84:743–756

Dewey C.E., Martin SW, Friendship RM, Wilson MR. The effects on litter size of previous lactation length and previous weaning-to-conception interval in Ontario swine, Preventive Veterinary Medicine 1994;183:213-223

Faccin J.E.G., Laskoski F., Hernig L. F., Kummer R., Lima G. F. R., Orlando U. A. D., Gonçalves M. A. D., Mellagi A. P. G., Ulguim R. R., and Bortolozzo F. P... Impact of increasing weaning age on pig performance and belly nosing prevalence in a commercial multisite production system. *J. Anim. Sci.* 2020b 98; 101093

Faccin J.E.G., Tokash M.D., Allerson M.W., Woodworth J.C., DeRouchey J.M., Dritz, S.S., Bortolozzo F.P. Goodband R.D. 2020. Relationship between weaning age and antibiotic usage on pig growth performance and mortality. J. Anim.Sci.98. 10. 1-10

Hamilton M.J., Frei S.M., Stevens R.L. The multifaceted mast cell in inflammatory bowel disease. Inflamm Bowel Dis. 2014; 20: 2364–2378.

Harris DL. Multisite PIG Production 2000. Iowa State University Press, Ames, Iowa USA.

Harris D. L., and Alexander T. J.. . 1999. Methods of disease control. In: Straw B. E., D'Allaire S., Mengling W. L., and Taylor D. J., editors. *Diseases of swine*. 8th ed. Ames (IA): Iowa State University Press; p. 1077–1110

Hu C.H., Xiao K., Luan ZS, Song J. Early weaning increases intestinal permeability, alters expression of cytokine and tight junction proteins, and activates mitogen-activated protein kinases in pigs. J Anim Sci. 2013; 91:1094–1101.

Le Cozier Y., Dagorn J., Dourmad J.Y., Johansen S., Aumaître A.. Effect of weaning-to-conception interval and lactation length on subsequent litter size in sows, Livestock Production Science, Volume 51, Issues 1–3,1997, Pages 1-11,

Main RG, Dritz S.S., Tokach M.D., Goodband R.D., Nelssen J.L. Increasing weaning age improves pig performance in a multisite production system. J Anim Sci. 2004;82:1499–1507

Main RG, Dritz S.S., Tokash M.D., Goodband R.D., Dhuyvetter K.C. Nelssen J.L. A partial-budgeting tool to describe the effect of lactation space and lactation space management on net revenue in the multi-site production system. J Swine Health and Prod. 2005; 13:322-332

Main RG, Dritz S.S., Tokach M.D., Goodband R.D., Nelssen J.L. Effect of weaning age on Growing-pig cost and revenue in a multi-site production system. J Swine Health and Prod. 2005; 13: 189-196

McLamb B.L., Gibson A.J., Overman E.L., Stahl C., Moeser A.J. Early weaning stress in pigs impairs innate mucosal immune responses to enterotoxigenic E. coli challenge and exacerbates intestinal injury and clinical disease. PLoS One. 2013;8:e59838.

Medland J.E., Pohl C.S., Edwards L.L., Frandsen S., Bagley K., Li Y. Early life adversity in piglets induced long-term upregulation of the enteric cholinergic nervous system and heightened sex-specific secretomotor neuron responses. Neurogastroenterol Motil. 2016; 9: 1317–1329.

Moeser A.J., Blikslager A.T. Mechanisms of porcine diarrheal disease. J Am Vet Med Assoc. 2007; 231: 56–67.

Moeser A.J., Klok C.V., Ryan K.A., Wooten J.G., Little D., Cook V.L. Stress signaling pathways activated by weaning mediate intestinal dysfunction in the pig. Am J Physiol Gastrointest Liver Physiol. 2007; 292: G173–G181

Moeser A.J., Ryan K.A., Nighot P.K., Blikslager A.T. Gastrointestinal dysfunction induced by early weaning is attenuated by delayed weaning and mast cell blockade in pigs. Am J Physiol Gastrointest Liver Physiol. 2007; 293: G413–G421.

Moeser A. J., Pohl C. S., and Rajput M M. Weaning stress and gastrointestinal barrier development: implications for lifelong gut health in pigs. *Anim. Nutr.* 2017: 3:313–321.

Pohl C.S., Medland J.E., Mackey E., Edwards L.L., Bagley K.D., DeWilde M.P. Early weaning stress induces chronic functional diarrhea, intestinal barrier defects, and increased mast cell activity in a porcine model of early life adversity. Neurogastroenterol Motil. 2017

Pohl C.S., Medland J.E., Moeser A.J. Early life stress origins of gastrointestinal disease: animal models, intestinal pathophysiology, and translational implications. Am J Physiol Gastrointest Liver Physiol. 2015; 309: G927–G941.

Smith F., Clark J.E., Overman B.L., Tozel C.C., Huang J.H., Rivier J.E. Early weaning stress impairs development of mucosal barrier function in the porcine intestine. Am J Physiol Gastrointest Liver Physiol. 2010; 298: G352–G363.

Smith A. L., Stalder K. J., Serenius T. V., Baas T. J., and Mabry J. W.. Effect of weaning age on nursery pig and sow reproductive performance. *J. Swine Health Prod.* 2008 ;16:131–137.

Advanced
Nutrition & In-Barn
Service.
FORTiFIED
NUTRITION
LIMITED
Grand Valley
FORTIFIERS
FOR SOUND NUTRITION
Canadian Family Owned

GET
BIG
FAST
DANBRED
DUROC
BRED FOR
BUSINESS
DANBRED
www.danbred.com
dave@danbred.com

The PigFlow® Point of Care Platform

Matthew Rooda

SwineTech, Inc., Solon, IA
mrooda@swinetechnologies.com, (641) 670-0774

Introduction

Significant losses have been attributed to sow and piglet mortalities. The global swine industry is continuing to see a rise in overall unexpected sow mortality at an average of ~15%, according to Chris Rademacher, DVM, extension swine veterinarian, Iowa State University. Additionally, 1.08 piglets per litter are found to be dead due to complications during the birthing process (stillborns). Once born, the first three days of the piglet's life represent the greatest danger of death. Approximately 70% of these piglet deaths occur during this time frame and according to the MetaFarms Production Index roughly 13.5% of all pigs born alive will not survive.

Furthermore, labor turnover and access to skilled labor are plaguing the US Swine Industry. The National Pork Board 2016-2017 "Employee Compensation & HR Practices in Pork Production" report shows the average reported turnover rate for animal caretakers on large and small to midsize producers was 35.03% and 20.34%. In many cases employee turnover will exceed 100%. This is a very costly problem for producers representing 4.05 billion U.S. dollars in annual losses. This high turnover rate is also resulting in an annual loss of on-site knowledge, experience, and skill sets. It is leading to greater variability in processes and pig care.

Our greatest challenge is that as an industry we are unaware and unable to measure what is being completed daily, who is completing it, and whether it was completed correctly as indicated by company SOPs. A solution is needed to bring more organization, communication, and visibility to daily operations so that farms can efficiently and effectively train new team members, improve workflow efficiencies, process compliance, and daily pig care.

Research and Methods

Industry interviews were conducted with 33 of the top 40 producers in the US, and 93.9% of these producers agreed that there is a significant lack of visibility into daily workflows, process compliance, and pig cares. In response to these interviews, 167 interviews were conducted with on farm personnel to understand the key stressors and issues that mattered to them. There were 103 English speaking participants and 64 Spanish speaking participants. Each participant was interviewed in their native tongue. The answers were consistent from farm to farm and fell into the following categories:

- Desire a more organized and timely onboarding experience
- Communication throughout the workplace is poor due to language barriers
- It is not clear throughout the day where people are and what needs to happen next

- Many tasks were inefficient in both preparation and execution
- The use of notepads causes information to remain invisible to the greater team
- Traditional handhelds offer no value to employees on the farm
- A strong desire to see KPIs tied to their individual performance

These responses indicated an opportunity to improve the quality of life of employees while also driving a more efficient workplace with improved pig cares. It was deemed that in order to properly address process compliance in sow farms a solution was needed for these employees that actually makes their quality of life better.

Results and Discussion

In response to these findings a group of 11 team members at SwineTech, Inc. and a couple dozen industry members which included owners, veterinarians, production supervisors, farm managers, and on farm employees collaborated in the development of the PigFlow Platform. Inspired by workforce management solutions in long-term healthcare facilities, PigFlow is a specialized communication tool that enables an optimized flow of information that delivers actionable intelligence to the person most able to act in a timely enough manner to achieve the best outcome at every level of business. It also provides a record of both the communication and the resulting action in order to evaluate the effectiveness of decisions and processes. This has led to improvements in labor time to productivity, pig survivability, and the implementation and management of precision technologies. PigFlow is currently managing ~350,000 sows in the US and Canada on sow farms from 400 sows to 12,000 sows. Producers have reported a reduction in labor needs, stillborns, pre-wean mortality, sow mortality, as well as a significant improvement in farrowing rates. In addition to these benefits, teams are more connected than ever and have reported to be more organized and less stressed throughout the day. It is our expectation that over time we will continue to see sustained improvements in employee onboarding, development, and retention.

In addition to the philosophy that drives PigFlow and its many outcomes, this presentation will also focus on new features that are being developed in collaboration with many reputable swine veterinarians, Johnsonville Foods, and Genus PIC to help producers accomplish the following:

- Management of Prop 12 operations.
- Custom treatment regimens for both medical and non-medical interventions.
- Measurement of treatment and vaccination compliance.

References

Ketchem, Ron. "70% Of Stillborns Come from 20% of the Sows; Who Is at Risk?" *National Hog Farmer*, 5 June 2019, https://www.nationalhogfarmer.com/reproduction/70-stillborns-come-20-sows-who-risk.

"**MetaFarms Production Index Analysis**." MetaFarms, 2020.

Wilson, Keith. "Sow Mortality Frustrates Experts." *National Hog Farmer*, 4 Feb. 2016, https://www.nationalhogfarmer.com/mag/farming_sow_mortality_frustrates.

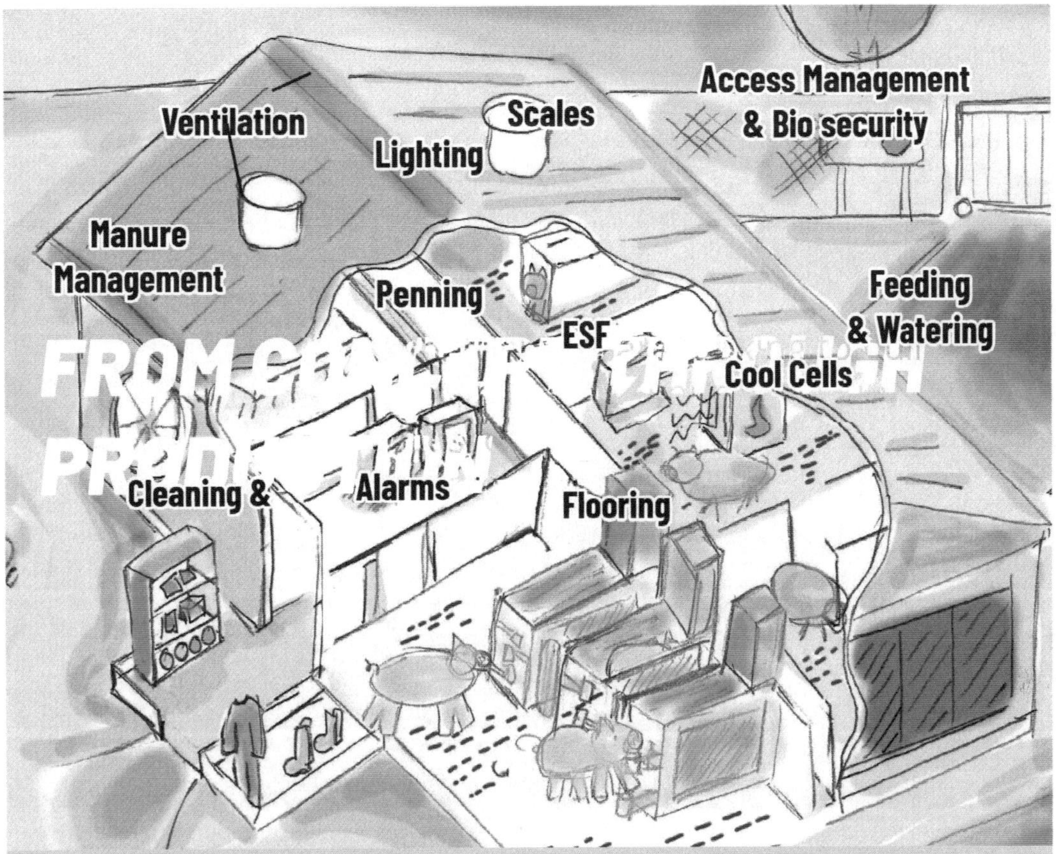

Ventilation
Scales
Lighting
Access Management
& Bio security
Manure
Management
Penning
Feeding
& Watering
ESF
Cool Cells
Cleaning &
Alarms
Flooring

AnimalHealthCanada
AHC
SantéAnimaleCanada
SAC
EMERGENCY MANAGEMENT DIVISION
DIVISION DE LA GESTION DES URGENCES
Create a plan
Élaborez un plan
Be prepared
Soyez préparé
Stay informed
Restez informé
animalhealth.ca

FEEDING SYSTEM
Gestal
FEED BLENDING
THE INDUSTRY-PREFERRED SOLUTION
CUSTOMIZABLE DIETS
Feed sows according to their needs, parity, days in lactation & piglets fed
REDUCE FEED WASTE
Minimize feed waste and costs by accurately distributing the right amount of feed to each sow
IMPROVE PERFORMANCE
Adjust your daily feeding curve to eliminate phase feeding & improve your herd management
JYGA
Technologies

ProHog Management: A Refresh on Margin

Steve Dziver

President, TrueAg Developments Inc
Email steve.dziver@trueag.ca, Ph: (204) 794-7757

1. Introduction

ProHog Management (ProHog) has been developed utilizing 20+ years of extensive analytical experience in the hog industry to support better agribusiness decisions tailored to meet the needs of the individual farm, integrated producer, corporate business, and industry service provider. ProHog Management is a complimentary add-on to any functioning herd management or data collection software providing real-time analysis of various economic and production influences.

The software is an innovative cloud-based service that integrates existing data with online sources in one place, allowing for improved collaborative abilities to save time and simplify decisions with customized reports and multi-factor scenario analyses.

Figure 1: Data Management

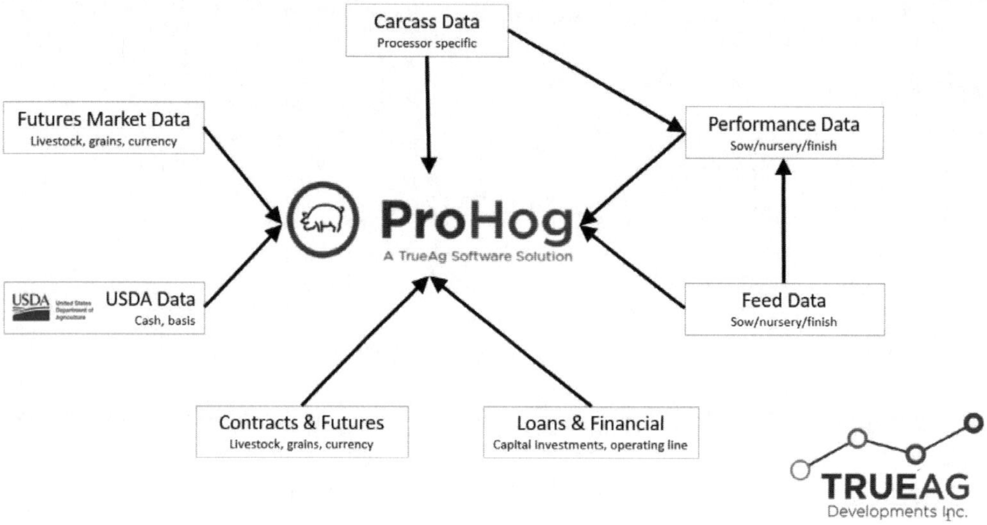

ProHog is a software that brings together livestock industry performance and economic data to a collective location so all users can maximize efficiencies and productivity, resulting in improved animal health and increased outputs including but not limited to economic improvement, key performance indicator tracking, cash flow analysis and projected margins.

2. Purpose

ProHog Management software has been developed to improve efficiencies in handling existing data by centralizing industry software platforms to support continuous improvement programs. In the realm of **multi-factor scenario analysis**, both producers and industry service providers contribute to the evaluation of sensitivity analysis, Return on Investment (ROI), and Key Performance Indicator (KPI) reporting. ProHog Management allows producers to analyze scenarios related to market fluctuations and production changes, aiming to optimize performance, meanwhile, providing industry service providers in areas such as feed, health, genetics, and equipment a tool to provide valuable data and insights to enhance decision-making within the industry.

Budgeting and forecasting in the agricultural sector involve critical considerations such as managing margins and cash flow, making productivity decisions, as well as strategically planning for expansion, re-population, or de-population. ProHog Management plays a pivotal role in this process by assessing market trends and aiding in decisions to enhance profitability. Simultaneously, industry service providers, including lenders, partners, and accountants, can collaborate with producers using ProHog, offering financial insight and strategic support.

Figure 2: Reporting

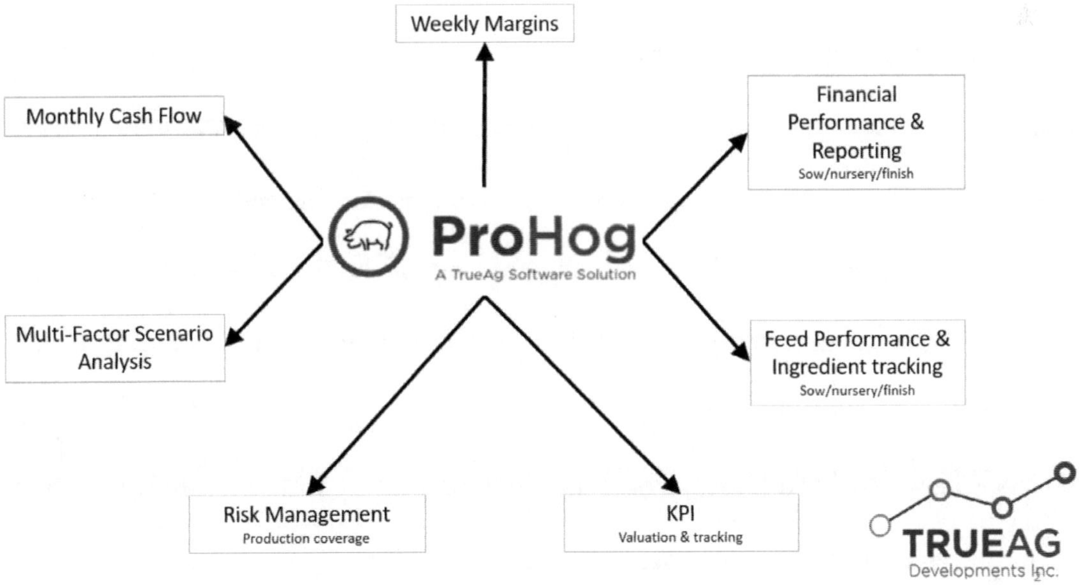

Lastly, effective **risk management** is integral to the success of agricultural businesses. ProHog Management allows producers to engage in market exposure calculations to assess and mitigate risks associated with market volatility and price fluctuations. It also brings together a common working platform for industry service providers, including marketers, brokers, and those involved in feed and commodity sales, to collaborate with producers to navigate and manage various risks effectively. ProHog Management promotes synergy between producers and industry service providers to encourage vital collaboration for the financial health and sustainable growth of this agricultural sector.

3. Functions

The software offers users a personalized online platform designed to deliver tailored commodity market updates based on specific agribusiness and location parameters. This platform enables users and authorized associates to instantly recreate and replicate existing farms, facilitating detailed margin recalculations and updated cash flow projections in response to production or cost changes. Service personnel, farm owners, or workers can independently assess single-factor analysis changes or conduct complex evaluations of multi-layered production and market variations, providing insights into economic and monetary output. The results of this instant analysis can be seamlessly integrated into farm strategies and complimentary workflow programs and software, with the ability to track changes over time. ProHog Management provides endless opportunities to accurately identify and quantify problem areas in production or marketing. Its economic and monetary reporting capabilities elevate decision-making by offering a comprehensive framework.

Notable features of the program are the ability to create, automate, and manage online Risk Management Plans. The software allows users to convert commodity base price units to their specific agribusiness units (e.g., per hog, per kg, per bushel, per tonne), facilitating easy analysis. Users can also delve into the analysis of the cost of goods purchased and sold relative to their agribusiness, evaluating the impact of these costs on margins and the overall risk management plan. Additionally, the software tracks forward sales and purchases for margin management purposes, and it records and reports the impact of customer-specific pricing decisions through automated tracking systems. Overall, this comprehensive platform empowers users in the agricultural sector to make informed decisions, optimize financial strategies, and adapt to dynamic market conditions.

SKOV
Climate for Growth
BlueControl
Breaks down the complexity
• Focus on productivity and animal welfare
• Timely and relevant information
• Complete toolbox
• Clear and simple overview
• Complexity is made simple
You control - BlueControl
www.skov.com

With a solid history of supporting companion animal and equine health, Dechra is continuing to extend its unique portfolio in livestock farming.
NEW!
Tri-Solfen™
Rexxolide®
Cyclospray
Dechra Veterinary Products Inc.
Tel.: 1-855-332-9334 | Technical Services: technical.ca@dechra.com | www.dechra.ca
Dechra

What is Genomic Selection and What Has Been Its Impact on Current Genetics

Tom A. Rathje[1,] Egbert F. Knol[2], Daniela Grossi[3], Patrick K Charagu[4] and Bob Kemp[5]

[1]DNA Genetics LLC, Columbus, NE 68434, U.S.A. email: trathje@dnaswinegenetics.com ; [2] Topigs Norsvin; [3]AcuFast; [4]Hypor; [5] Genesus ; All companies are members of PigGen Canada (https://piggencanada.org/).

1. Introduction

Genomics is a broad term that describes the use of multiple technologies used to gain an understanding of the variation in DNA sequence and how that variation relates to, in the case of pigs, animal performance. To date, the most impactful implementation of genomic technology in modern pig breeding has been the use of a group of genetic markers known as SNP (single nucleotide polymorphism) to implement genomic selection. SNP are markers present in the pig genome that represent a location in the pig's genome that contains a single DNA base-pair that varies within pig populations. They are found in the genome by sequencing the DNA of a sample of pigs within and across breeds. While the total number of SNP found within the genome of pig is in the many millions, a subset of these SNP (50-70,000) is selected to produce a genotyping array. These selected SNP highlight differences in the genetic make-up of individual pigs across the entire pig genome as they are relatively evenly distributed across the pig genome. For most of the SNP, we do not know their biological function. For others, we do, and that is the subject of a second paper in this section.

The process of applying genomic selection begins with taking a tissue, blood or semen sample from pigs that are candidates for selection. The DNA is extracted from this sample and is further analyzed to produce the genotype of the pig at each of the individual SNP present on the genotyping array. This results in a data file containing the genotypes which can then be used for multiple applications. This paper will focus on the use of these data for genomic selection in pigs.

2. What is Genomic Selection

To understand the impact of genomic selection, a bit of background on a genetic analysis to predict breeding values is helpful. The genes that affect a particular trait can take on various forms called alleles. In pig breeding our goal is to predict which animal is most likely to possess a group of alleles which positively affect the traits we are interested in. We express this prediction as an estimated breeding value (EBV). An EBV can be thought of as a single number representing the cumulative (additive) effect of all alleles impacting a trait of interest. Historically, this is accomplished by measuring the performance of each pig (e.g. growth rate, feed efficiency, litter size) within a contemporary group. These data are adjusted to remove known sources of variation in performance such as sex, season of the year, barn or room and pen effects to produce an environmentally adjusted phenotype for comparison. These adjusted phenotypes are correlated among relatives (animals sharing alleles) from common ancestors) and this allows us to utilize a prediction method that makes use of information from relatives to increase the accuracy of an EBV for any given animal. The performance of an animal compared to their contemporary group helps predict the performance one could expect from a relative sharing a similar genetic profile or make-up.

Prior to genomic selection, the prediction model used relationships between animals that are based on the pedigree and take on values such as 0.5 as the relationship between parent and offspring or between full-sibs or 0.25 between half-sibs and between an animal and a grandparent. Animals not connected through pedigree are considered unrelated. Despite this generalization of relationships, the method effectively modeled the data structure of livestock performance trait measures and led to effective and accurate responses to selection (Goddard et al., 2010).

Genomic selection differs specifically in the calculation of the relationships between individuals. As an example, in a litter of pigs that are full-sibs, each pig shares on average half its DNA with a sibling. This is the value of 0.5 referenced earlier. However, the reality is that individuals that are full-sibs can share anywhere from 0 to 100% of their autosomal DNA, having received a random sample of the DNA from each parent. The use of SNP markers distributed across the genome allows a prediction of the 'true' relationship between any two individuals. Since the relationship is calculated at the DNA-level, this approach also calculates relationships between previously assumed unrelated individuals by examining the SNP markers in common between them. The use of genomic data in this way greatly enhances the accuracy of an individual's breeding value (now called a genomic estimated breeding value, gEBV) because one has more precisely weighted the data available from relatives and non-relatives sharing common DNA. The increased accuracy of a gEBV results in a higher genetic trend for the traits of interest (Knol et al., 2016), which is the outcome of genomic selection most relevant to pork producers.

There are additional benefits to the use of genomic data beyond the improved accuracy of the gEBV. For example, genomic relationships between animals can be used to verify parentage or to assign parentage to an individual via comparison of a genotype to a group of potential parents. This is particularly useful when working with crossbred data derived from commercial production systems. Somewhat more complicated, it is possible to back-calculate the effects of the individual SNP markers (or groups of markers) and thereby shed light on the areas of the pig genome that associate with performance for a trait of interest. This latter application can assist with identifying the causative allele, which has relevance for things like disease resistance wherein we may choose to place direct selection pressure in favor of a particular allele conferring resistance to health challenges (e.g. E.coli.F18, WUR, HAL). Further discussion of these applications is beyond the scope of this paper.

3. What has been the impact of Genomic Selection on Current Genetics

All major genetic suppliers to the pork industry today utilize genomic selection in their breeding programs at various levels. This has resulted in substantial increases in the genetic potential available to producers over the past decade when compared to any prior decade. The impact of gEBV on selection response varies by trait with the largest impact being for those traits that are of the lowest heritability (e.g. mortality, litter size), are measured in a single sex (e.g. litter size) or are measured late in life (e.g. sow retention). The paper by Knol et al. (2016) provides examples of the impact of genomic selection on the accuracy of predicting breeding values in pigs. The accuracy of breeding value for teat count, which has a high heritability of 0.4, increased by 50% with the use of genomic selection. Post-weaning mortality, which has a very low heritability of 0.05, also resulted in an increase in accuracy of 50%. In a third example using an overall index trend (with 8 traits) demonstrated an increase in progress of 50% when comparing the trend before and after the implementation of genomic selection. Other real-world examples will be shared during the break-out session which support the impact of genomic selection on overall economic value transferred to the pig producer due to superior genetic gain.

4. Implications

Genomic selection has increased the accuracy by which we can identify genetically superior nucleus animals and results in a direct economic benefit to the pork producer, increasing their return on investment in genetics. The use of genomic selection has also increased genetic change in lowly heritable and difficult to measure traits, including traits not easily measured in a high-health nucleus system such as sow retention and pre- and post-weaning mortality. The use of genomic technology will continue to expand as we continue exploring new methods to further optimize and improve the efficiency of breeding programs, incorporate commercial-level crossbred data and to continue address issues of sustainability for the pork industry.

References

Goddard, M.E., B.J. Hayes and T.H.E. Meuwissen. 2010. Genomic selection in livestock populations. Genet. Res., Camb. 92:413-421

Knol, E.F., B. Nielsen and P.W. Knap. 2016. Genomic selection in commercial pig breeding. Animal Frontiers. 6(1):15-22

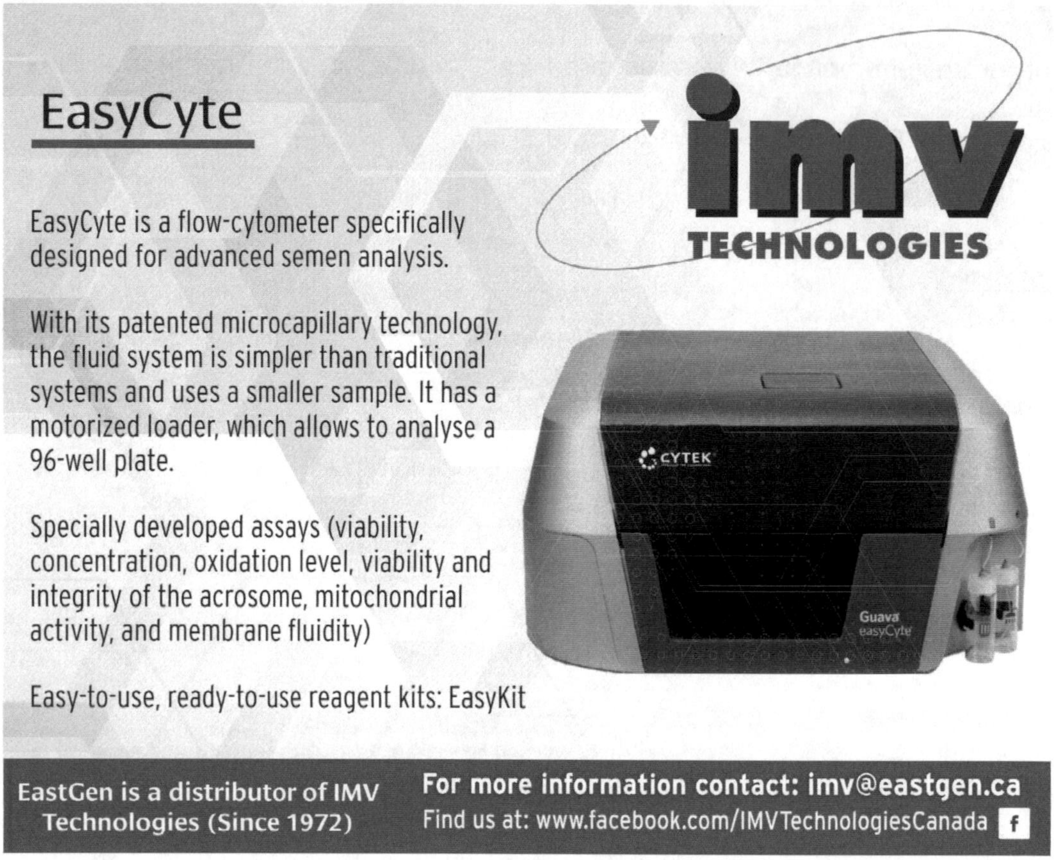
EasyCyte
EasyCyte is a flow-cytometer specifically designed for advanced semen analysis.
With its patented microcapillary technology, the fluid system is simpler than traditional systems and uses a smaller sample. It has a motorized loader, which allows to analyse a 96-well plate.
Specially developed assays (viability, concentration, oxidation level, viability and integrity of the acrosome, mitochondrial activity, and membrane fluidity)
Easy-to-use, ready-to-use reagent kits: EasyKit
imv TECHNOLOGIES
CYTEK
Guava easyCyte
EastGen is a distributor of IMV Technologies (Since 1972)
For more information contact: imv@eastgen.ca
Find us at: www.facebook.com/IMVTechnologiesCanada

Boehringer Ingelheim
COMPLETE CONFIDENCE IN OUR CUSTOMIZED PORTFOLIO
WITH COMPLETE HEALTH SOLUTIONS COMES SOMETHING MUCH GREATER: COMPLETE CONFIDENCE.
ENTERISOL SALMONELLA T/C®, INGELVAC®, INGELVAC CIRCOFLEX® and INGELVAC MYCOFLEX® are registered trademarks of Boehringer Ingelheim Vetmedica GmbH, used under license.
PARASAIL® is a registered trademark of Boehringer Ingelheim Animal Health USA, Inc., used under license.
ENTERISOL™ is a trademark of Boehringer Ingelheim Vetmedica GmbH, used under license.
©2023 Boehringer Ingelheim Animal Health Canada Inc. All Rights Reserved. CA-POR-0017-2023
Visit SWINEINFO.CA to learn more

HyLife
HyLifepork
THE TASTE SPEAKS FOR ITSELF
Taste & Trust
THE HYLIFE DIFFERENCE
THE HYLIFE DIFFERENCE
HyLife's vision is to be the best food company in the world. Our fully integrated operations model sets us apart, allowing us to consistently create excellent products that are enjoyed around the globe. Our unique feeding program and the marbling of our Duroc breed hogs ensure unmatched flavour.
Animal Genetics
Feed Production
Animal Care
Transportation & Logistics
Processing & Spec
Global Distribution

Individual SNP Effects and Interesting Applications in Pig Breeding

Egbert F. Knol[1], Tom A. Rathje[2], Daniela Grossi[3], Patrick K. Charagu[4], and Bob Kemp[5]

[1]Topigs Norsvin, Den Bosch, the Netherlands, Egbert.Knol@TopigsNorsvin.com
[2]DNA Genetics; [3]AcuFast; [4]Hypor; [5]Genesus.
All companies are member of PigGen Canada (https://piggencanada.org/)

Introduction

In general genomic selection helps in traits which are (1) late in life, (2) sex limited, (3) difficult to measure and (4), lowly heritable. As an example: litter size, this can be measured only in females and only when the animal IS already selected (so late in life). Difficult to measure: carcass composition and meat quality; this can be done after slaughter, rendering the animal unfit for reproduction. Prime example of a trait NOT in need of genomic selection is #teats. This can be counted early in life (day of birth), on both sexes, most people can count to 25 and has a heritability of 40%.

A SNP is a Single Nucleotide Polymorphism and can serve as a marker for a functional mutation, that is, a change in the DNA that results in a change in performance, often a protein that is not produced anymore.

Individual SNP Effects

Rathje et al. (2023) explained the relationship matrix approach to genomic selection. Animals that have many SNP's in common will show more similar phenotypic performance. Proper statistical analysis can give estimates for the individual SNP's. This estimate is then an indication of how relevant the functional mutation is.

Another approach is to run an analysis in isolation. If there is a specific problem or disease or new trait of interest then observations can be taken, the same animals can be genotyped and a Genome Wide Association Study (GWAS) can be run. A proper GWAS needs, most of the times, 1000 observations or preferably more. Figure 1 gives a GWAS of # teats. On the X-axis are the 18 autosomes. The Y-axis gives the significance of a specific SNP, expressed as a log (P-value). A value of 60 indicates that the change that the SNP is not important is less than 0.00..001, where there are 60 zeros behind the decimal. This specific plot shows that there is a mutation on chromosome 7 which highly significantly influences # teats. Since this SNP appears in all three lines at exactly the same location it might very well be an old mutation, which came about in pigs before they separated into individual breeds or lines.

This analysis of #teats is based on imputed 660k SNP's (as explained by Rathje et al., 2023), and allows for a search for the functional mutation. This was done by van Son et al. (2019) (graphs are from this paper), they found a mutation that causes an increase in the #vertebrae, allowing the growth of 2 extra ribs, which then allow for the formation of two extra teats.

Applications

Polygenic approach: Estimating the effects of SNP's allows for the selection of animals which to not (yet) have own observations. Analysing the DNA of these animals and using the SNP estimates gives a proper estimation of the breeding values of these animals. If the functional mutations behind these SNP effects are known, then the accuracy, the certainty of the effect increases greatly. A word of caution: a lot of mutations occur within a line or breed, if that is the case, the effect cannot be used in another line or breed.

Marker assisted approach: There are SNP's with a large effect (eg # teats). Aside of relying on the relationship matrix one could also use the SNP explicitly in the index, only keep the desired allele. Theory has gradually turned against this approach, since there might also be negative effects of the allele at stake (negative pleiotropy) or there might be genes nearby which automatically are selected too. In general it increases inbreeding.

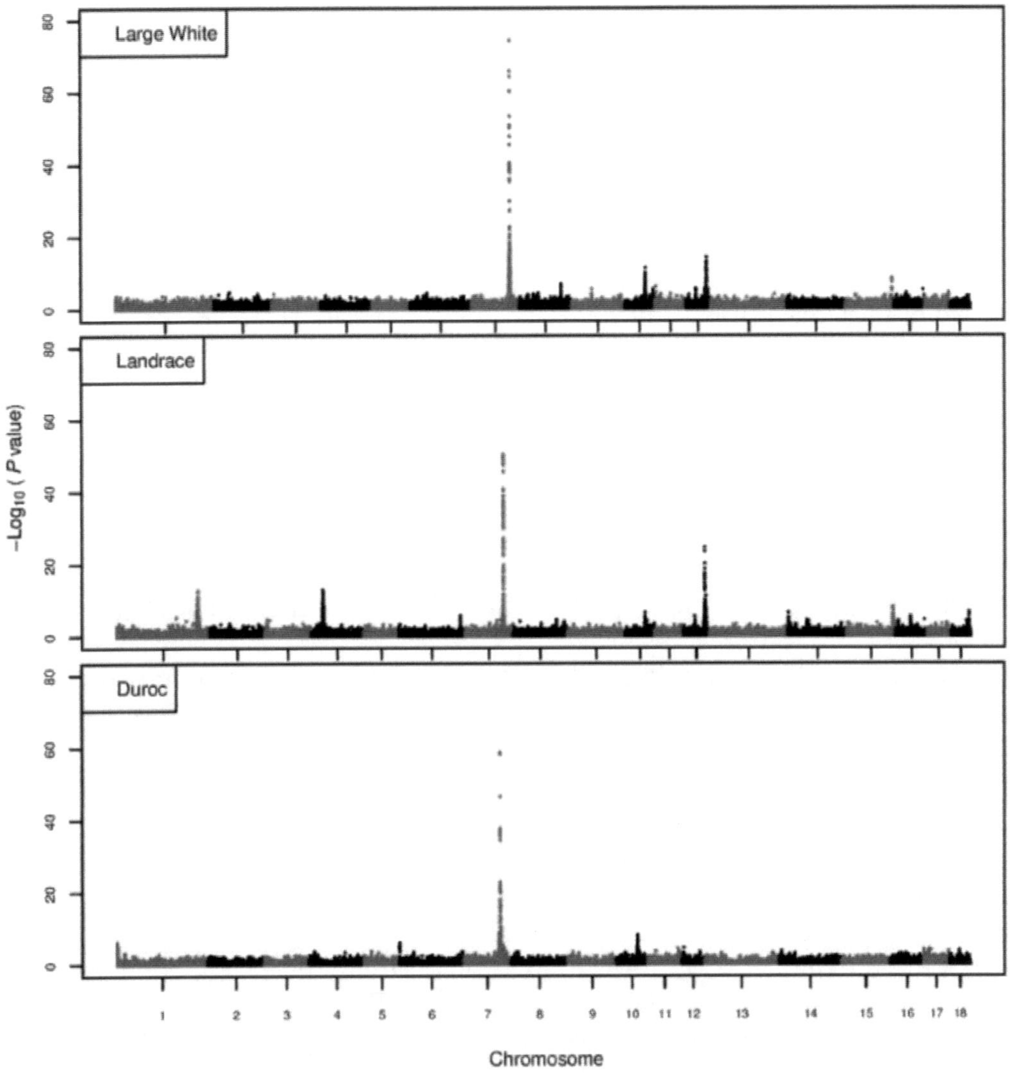

Figure 1: GWAS plot for NTE with imputed 660K SNP chip data. On the *y*-axis is the $-\log_{10}(p\text{-values})$ of single SNP association with NTE in LW, L, and D breeds. On the *x*-axis is the physical position of the SNP across the 18 autosomes.

Gene editing targets: Assume a positive mutation exists in line A and not in Line B and the mutation involves a single or a few SNP's, then this mutation can be created in line B instead of multigenerational linecrossing.

Single Genes and Disease

Plant breeders are quite successful in selection against disease. Pathogens adapt to their hosts and, among others, can find a place of entry, through eg a receptor on the cell surface. This approach works very well until the pathogen has changed enough to be infectious again.

This is the case with PRRS. Its sole way of entry is through a receptor called CD163. A simple edit can make the receptor impossible to attach to, rendering it impossible to make the animal sick. One of the issues here is that almost all edits work through a recessive model: both chromosomes of an animal have to carry the allele that cripples the receptor, if one of the chromosomes still carries the original version then the disease can still enter. For crossbred finisher pigs this means that 3 populations have to be homozygous recessive, a huge effort in animal selection.

Piggen Canada has just finished a very large program on health and resilience where one of the findings was a mutation in GPB5. This gene produces a protein that helps to fight virus attacks. A mutation has crippled this gene in some families, but genomic selection will help to increase the original version of the gene in the population (Jeon et al. 2021). This is a dominant approach. Only one of the chromosomes has to have the positive allele to be active, to fight the pathogen. This allows for homozygous positive boars to have an immediate very wide influence.

Implications

This technology is developing very quickly and makes use of knowledge in different species and growing technology in genotyping and sequencing technology. Rate

AND accuracy of genetic improvement will further increase and will be used more and more in difficult traits as health and behavior.

References

Derks, M. et al. 2021. A natural knockout of the MYO7A gene leads to pre-weaning mortality in pigs. Anim. Genet. 52:514–517.

van Son M et al. 2019. A QTL for Number of Teats Shows Breed Specific Effects on Number of Vertebrae in Pigs: Bridging the Gap Between Molecular and Quantitative Genetics. Front Genet. 2019 https://doi:10.3389/fgene.2019.00272. PMID: 30972109; PMCID: PMC6445065.

Jeon et al. Effect of a genetic marker for the GBP5 gene on resilience to a polymicrobial natural disease challenge in pigs https://doi.org/10.1016/j.livsci.2021.104399

BIRTH TO PLATE
Ketchum ID Traceability products have you covered.
Canadian solutions since 1913.
B7 SHOULDER MARKING INK
BLACK
NOIR
ENCRE B7 D'ABATTAGE
Since 1913
Ketchum
Manufacturing Inc.
1245 California Avenue, Brockville, Ontario, Canada K6V 7N5
www.ketchum.ca | 613-342-8455

Because you can't make split-second decisions with old data.
Build a better pork system
Our Producer Profitability Tool enables you to make the right decisions at the right time.
Reach out to your local sales consultant to learn more.
Cargill
©2023 Cargill, Incorporated. All Rights Reserved.

Indicators and Genetics of Disease Resilience

Jack Dekkers[1], Fred Fortin[2], Mike Dyck[3], John Harding[4], Yolande M. Seddon[4], Graham Plastow[3], PigGen Canada[5]

[1]Department of Animal Science, Iowa State University, 806 Stange Rd, Ames, IA, 50011. Email: jdekkers@iastate.edu; [2]Centre de développement du porc du Québec inc., Québec City, QC; [3] Department of Agricultural, Food and Nutritional Science, University of Alberta, Edmonton, AB; [4] Western College of Veterinary Medicine, University of Saskatchewan, [5]PigGen Canada, Guelph, ON.

1. Introduction

Infectious disease represents one of the largest costs of pork production, leading to veterinary costs, reduced performance, morbidity, mortality, and increased labor. Disease also affects animal welfare and morale of farm workers. The incidence and impact of infectious disease can be reduced by implementing biosecurity protocols, vaccination, and routine therapeutic or metaphylactic use of medications. Susceptibility to most infectious diseases is in part determined by the genetics of the animal, as is how the animal responds to an infection. Thus, genetic selection is another strategy that can be used to reduce disease and/or the impact of disease. Unfortunately, the ability to make animals completely resistant is limited to only a few pathogens (e.g. F18 E.coli) or is only possible by gene editing (e.g. the PRRS-resistant pig created by editing the CD163 gene). For most other infectious diseases, there are genetic differences in the degree of resistance or susceptibility between animals. This implies that genetic selection should not be used as the only strategy to combat disease, but nor should other strategies because biosecurity protocols that keep most/all infectious pathogens out are not feasible at the commercial level in hog-dense regions and effective vaccines are only available and/or efficacious for some pathogens. Given these limitations, commercial pigs will continue to be exposed to, and infected by, pathogens for the foreseeable future.

1.1 Disease Resilience

Disease resilience is the ability of an animal to be minimally affected by disease or to rapidly return to normal performance after pathogenic infection. Combined with the fact that pathogens evolve, that new (foreign) diseases can emerge (e.g. African Swine Fever) and that complete resistance is not possible for most pathogenic diseases, disease resilience is a promising target for inclusion in breeding programs. Disease resilience can be evaluated by exposing animals to disease and recording disease incidence, mortality, and performance. This is, however, not possible in the purebred nucleus herds that are the basis of modern pig breeding programs, because these herds must maintain a high health status. An alternative is to develop so-called genetic indicator traits for disease resilience that can be measured in the nucleus. A possible example of such a genetic indicator trait is a measure of an animal's immune response to a vaccine.

For an indicator trait to be useful to select for disease resilience, it must meet the following three criteria:

1) Be measurable on healthy animals without exposure to disease (i.e. in the nucleus), ideally at a young age.

2) Have at least moderate heritability, such that selection on it results in genetic change.
3) Have a sizeable genetic correlation with disease resilience, such that genetic change in the indicator trait results in genetic improvement of disease resilience as a correlated response.

The purpose of this presentation is to describe a large-scale study on the genetics of disease resilience that was conducted from 2015 to 2022 at the Center de Développement du Porc du Québec (CDPQ), in collaboration with PigGen Canada, and its use to identify possible genetic indicator traits.

2. The Natural Disease Challenge Model

The first step of this project was to establish a system and protocols to measure disease resilience on a large number of animals. For this purpose, a nursery-finish barn at CDPQ was seeded with multiple pathogens by bringing naturally infected pigs from strategically selected farms into the barn, targeting multiple pathogens, including several strains of PRRSV and swine influenza virus, *Mycoplasma hyopneumoniae* (M hyo), and *Lawsonia intracellularis*, as well as several opportunistic bacteria, including *Streptococcus suis, Salmonella spp.* and *Glasserella parasuis*. The aim was to mimic the environment of a commercial farm with high disease pressure to maximize expression of genetic differences in disease resilience. To maintain disease pressure over time without having to keep introducing pathogens, a continuous flow system was implemented, in which a new batch of naïve pigs entered the challenge nursery every three weeks at ~40 days of age, with fence-line contact with the preceding batch for ~1-week. Each batch consisted of 60 or 75 Large White x Landrace barrows from a high-health multiplier farm from one of the seven member companies of PigGen Canada. With careful veterinary oversight, the NDCM was maintained for seven years, exposing approximately 4,700 nursery pigs to this natural polymicrobial disease challenge across 68 batches (see Figure 1).

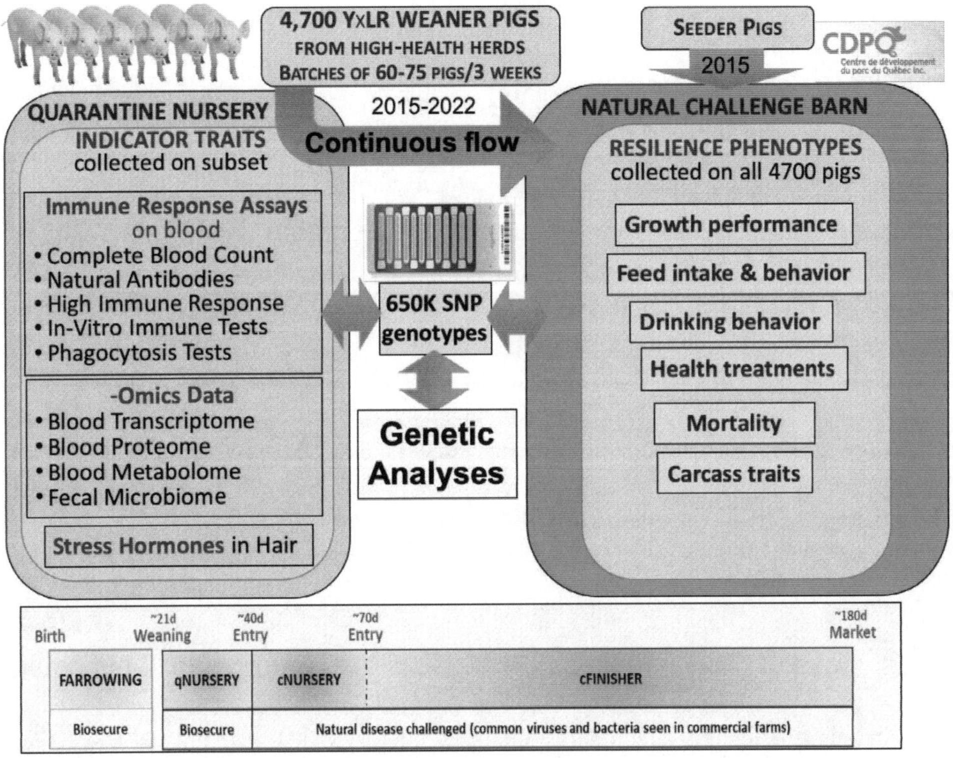

Figure 1. Overview of the Natural Disease Challenge Model (NDCM).

3. Measures and Genetics of Disease Resilience

The following phenotypes were recorded on all animals in the challenge barn of the NDCM as measures of disease resilience (Figure 1): growth rate, mortality, and all veterinary treatments in the nursery and in the finisher, feed intake (using electronic feeders) and FCR in the finisher, and carcass traits for pigs that made it to market. Drinking behavior was also recorded in the finisher. It should be noted that the purpose of the NDCM was not to compare disease resilience between batches, which was, in fact, not possible because disease pressure was dynamic over time. Instead, the purpose of the NDCM was to analyze differences in disease resilience between animals within a batch and to determine the extent to which these differences were determined by genetics (i.e. heritability). To enable genetic analyses, all pigs were genotyped for 600,000 genetic markers across the genome to quantify genetic relationships between pigs.

Most traits collected in the NDCM were found to be heritable (Putz et al. 2019, Cheng et al. 2020), indicating that a pig's resilience to polymicrobial disease is heritable and can be improved by selection. These heritabilities were despite the dynamic and relatively uncontrolled disease pressure in the NDCM, indicating that disease resilience has some common genetic basis across diseases.

One of the clinical signs of most infectious diseases is loss of appetite. This could be directly observed in the NDCM using data on individual feed intake in the finisher. To capitalize on this, the individual feed intake data were used to develop several novel measures of disease resilience based on day-to-day fluctuations in feed intake and the number of 'off-feed days' (Putz et al., 2019). Pigs with greater day-to-day variation in feed intake and more 'off-feed days' were found to have higher treatment rates and mortality, indicating that they are less resilient. These novel phenotypes derived from feed intake patterns were also moderately heritable and had moderately high positive genetic correlations with treatment rate and mortality, indicating that they could serve as genetic indicator traits to select for disease resilience (Putz et al., 2019; Cheng et al., 2020). Similar studies were conducted on drinking behavior (Cheng et al., 2021). In later batches, electronic feeders were also installed in the challenge nursery, where pigs were initially exposed, and the resulting feed intake data was also used to evaluate several novel disease resilience measures in the challenge nursery.

Each animal's genotypes for the 600,000 markers across the genome were also used to conduct genome-wide association studies on disease resilience phenotypes to identify genetic markers or regions of the genome that are associated with disease resilience (Cheng et al., 2022). These analyses showed that disease resilience is controlled by many genes across the genome but with a preponderance of genes associated with innate and adaptive immune response. In particular, the association of the Major Histocompatibility Complex (MHC) region with growth rate under challenge stood out. The MHC has important roles in immune response and this region has not been found to be associated with growth rate under healthy conditions.

4. Potential Genetic Indicator Traits for Disease Resilience

Prior to their entry in the challenge nursery, at ~40 days of age, each batch of pigs was kept in a quarantine nursery for ~18 days, after their arrival from a high-health multiplier farm (Figure 1). During this quarantine period, multiple measures and samples (blood, hair) were collected for the purposes of developing potential genetic indicator traits for disease resilience that could be measured on young healthy pigs. These included complete blood counts (Bai et al., 2020, 2021, 2022), multiple immune response assays measured on blood to characterize different components of the immune system of the animal (Chen et al., 2020; Jeon et al., 2021; Bai et al., 2022; Bhatia et al., 2022), the concentration of multiple metabolites (metabolomics) (Dervishi et al., 2021, 2023) and proteins (proteomics) (Chen et al., 2023) in blood, and the level of the expression of over 14,000 genes in blood (transcriptomics) (Lim et al., 2021, 2022). Because of the substantial costs associated with many of these assays, they were conducted on subsets of pigs (generally ~900) that subsequently entered the disease challenge. Resulting data were used to estimate the heritability of each assay and their genetic correlations with disease resilience phenotypes. Results showed some to have sufficient heritability and genetic correlations with resilience to have the potential to serve as genetic indicator traits that can be evaluated on young animals in the nucleus. However, because of the still limited number of animals with data on these assays, standard errors of the estimates of genetic correlations were large. Thus, additional validation of the most promising assays will be needed before they can be applied. Results from two recent examples are described next.

4.1 Feed intake patterns in nucleus herds

As described before, day-to-day patterns of feed intake and duration measured in the NDCM finisher are potential genetic indicators of disease resilience. These results, however, referred to feed intake patterns measured under disease. Individual feed intake data are routinely collected on purebred grow-finish pigs in nucleus herds. Therefore, the question is whether similar patterns measured in such biosecure environments are also genetic indicator traits for disease resilience. Although pigs in nucleus herds are not exposed to major pathogens, they are exposed to multiple non-pathogenic stressors that can result in fluctuations in feed intake, including mixing, diet changes, temperature, and handling. To investigate this, feed intake data on the parental lines from two companies that contributed LWxLR crossbreds to the NDCM were analyzed. Although day-to-day feed intake patterns exhibited similar heritabilities in these biosecure environments as in the NDCM, disappointingly most genetic correlations with disease resilience traits in the NDCM were fairly close to zero, but with large standard errors. More data is required to improve the estimates and determine if this approach could work.

4.2 Stress hormones in hair

In ongoing work, stress hormone levels in hair are investigated as potential genetic indicator traits for disease resilience. Stress hormones are deposited into hair growth at a concentration that reflects their concentration in blood. Thus, the concentration of stress hormones in hair reflects the animal's response to stress during the time the hair was grown. For this purpose, hair was shaved from the back of each pig in 15 batches just prior to their entry in the challenge nursery. Hormone levels in hair at that point reflects the animal's response to the stresses that it was exposed to up to that point in time, including weaning, mixing, transportation, and adaptation to new environments and diets. Although pigs are not expected to have been exposed to major pathogens by that point, given the cross-talk between immune and endocrine systems, the response to stress is expected to be related to the response to disease. To date, heritabilities of the concentration of two stress hormones, cortisol and DHEA-S, have shown moderately high heritabilities. Estimation of genetic correlations with subsequent disease resilience is underway.

5. Genomic Prediction of Disease Resilience

Another potential strategy is to use the disease resilience data from the NDCM to aid selection for disease resilience in purebred nucleus herds via genomic prediction. This approach uses genomic relationships between the LWxLR crossbreds evaluated in the NDCM and animals from their parental lines. To investigate the potential of this strategy, genomic prediction analyses were conducted within the NDCM, separately for the pigs entered from each company. Results showed reasonable accuracies of genomic prediction, given the limited number of pigs with NDCM data for each company (on average 500 in these analyses). Although genomic relationships between pigs from different companies were low (although belonging to the same breeds, their purebred lines have been separated from each other many generations), including data from the 6 other companies to derive genomic predictions for the 7th company increased the accuracy of genomic predictions. This demonstrates the value of companies sharing data for these difficult and costly to measure phenotypes.

6. Conclusions and Implications

Through the NDCM, unique and extensive data has been collected on a large number of wean-finish pigs that can be used to understand the genetic basis of disease resilience. Traits related to disease resilience were found to be heritable, which indicates that selection for disease resilience is possible, if suitable tools (genetic tests, genomic predictions, indicator traits) can be identified that can be implemented in nucleus breeding programs. To be useful, genetic indicator traits must be heritable and genetically correlated with disease resilience. Resilience measures derived from individual feed intake patterns in the finisher of the NDCM had potentially useful genetic correlations with treatment rate and mortality. Unfortunately, similar measures derived from feed intake data collected in biosecure nucleus herds did not show strong genetic correlations with disease resilience but additional research is warranted.

Assays conducted on blood from young healthy pigs showed promise as genetic indicator traits for disease resilience, although standard errors of the estimates tended to be sizeable. Thus, further work is needed to validate these findings and to implement them in nucleus breeding programs. Concentrations of the stress hormones cortisol and DHEA-S in hair of young healthy pigs were also heritable and have potential as genetic indicators for disease resilience. When implementing any genetic indicator trait, care must be taken that selection on it does not have unfavorable effects on other traits of interest.

Genomic prediction provides a valuable tool to predict breeding values on animals for traits for which they are not being recorded. Disease resilience is a good candidate for this. Genomic prediction, however, requires ongoing recording of the phenotype in a so-called training population. The NDCM can provide such data but continuous operation of the NDCM is costly. Collecting data from commercial herds that have a history of disease challenges would be an alternative that should be explored, now that the potential for genetic improvement of disease resilience has been demonstrated.

References

Bai, X., Putz, A.M., Wang, Z., Fortin, F., Harding, J., Dyck, M.K., Dekkers, J., Field, C.J. and Plastow, G.S., 2020. Exploring Phenotypes for Disease Resilience in Pigs Using Complete Blood Count Data From a Natural Disease Challenge Model. Frontiers in genetics, 11, p.216. doi.org/10.3389/fgene.2020.00216 [9]

Bai, X., Yang, T., Putz, A.M., Wang, Z., Li, C., Fortin, F., Harding, J.C., Dyck, M.K., Dekkers, J.C., Field, C.J. and Plastow, G.S., 2021. Investigating the genetic architecture of disease resilience in pigs by genome-wide association studies of complete blood count traits collected from a natural disease challenge model. BMC genomics, 22(1), pp.1-15. https://doi.org/10.1186/s12864-021-07835-4 [2]

Bai, X.; Cheng, J.; Fortin, F.; Harding, J.C.S.; Dyck, M.K.; Dekkers, J.C.M.; Field, C.J.; Rogel-Gailard, C.; Blanc, F.; Plastow, G.S. 2022. Indicators of disease resilience from complete blood count and in vitro immunoassays data from young-healthy pigs. Proceedings 12th World Congress on Genetics Applied to Livestock Production, the Netherlands. Pp. 696-699. https://doi.org/10.3920/978-90-8686-940-4_161

Bhatia, V.; Schmied, J.; Cheng, J.; Bai, X.; Mallard, B.; Fortin, F.; Harding, J.C.S.; Dyck, M.K.; Plastow, G.S.; Field, C.J.; Rogel-Gaillard, C.; Blanc, F.; Piggen Canada; Dekkers, J.C.M. 2022. Genetic relationships among immune response traits of young healthy pigs evaluated by immunoassays. Proceedings 12th World Congress on Genetics Applied to Livestock Production, the Netherlands. Pp. 3282-3285. https://doi.org/10.3920/978-90-8686-940-4_797

Chen, Y., Lonergan, S., Lim, K.S., Cheng, J., Putz, A.M., Dyck, M.K., Canada, P., Fortin, F., Harding, J.C., Plastow, G.S. and Dekkers, J.C., 2023. Plasma protein levels of young healthy pigs as indicators of disease resilience. Journal of Animal Science, 101, p.skad014. https://doi.org/10.1093/jas/skad014

Chen, Y., Tibbs-Cortes, L.E.T., Ashley, C., Putz, A.M., Lim, K.S., Dyck, M.K., Fortin, F., Plastow, G.S., Dekkers, J.C. and Harding, J.C., 2020. The genetic basis of natural antibody titers of young healthy pigs and relationships with disease resilience. BMC genomics, 21(1), pp.1-17. doi.org/10.1186/s12864-020-06994-0 [14]

Cheng J., R. Fernando, S. Kachman, H. Cheng, K. Lim, J. C. S. Harding, M. K. Dyck, F. Fortin, G. S. Plastow, PigGen Canada, and J. C. M. Dekkers. 2022. Genome-wide association studies on disease resilience traits from a natural polymicrobial disease challenge model in pigs identify the importance of major histocompatility complex region. G3, 12(3), p.jkab441. https://doi.org/10.1093/g3journal/jkab441 [2]

Cheng, J., Putz, A.M., Harding, J., Dyck, M.K., Fortin, F., Plastow, G.S. and Dekkers, J.C.M., 2021. Genetic parameters of drinking and feeding traits of wean-to-finish pigs under a polymicrobial natural disease challenge. Journal of animal science and biotechnology, 12(1), pp.1-19. https://doi.org/10.1186/s40104-021-00622-x [2]

Cheng, J., Putz, A.M., Harding, J.C., Dyck, M.K., Fortin, F., Plastow, G.S., Canada, P. and Dekkers, J.C., 2020. Genetic analysis of disease resilience in wean-to-finish pigs from a natural disease challenge model. Journal of Animal Science, 98(8), p.skaa244. doi.org/10.1093/jas/skaa244

Dervishi, E., Bai, X., Cheng, J., Fortin, F., Dyck, M.K., Harding, J.C., Seddon, Y.M., Dekkers, J.C., Canada, P. and Plastow, G., 2023. Exploration of plasma metabolite levels in healthy nursery pigs in response to environmental enrichment and disease resilience. Journal of Animal Science, 101, p.skad033. https://doi.org/10.1093/jas/skad033

Dervishi, E., Yang, T., Dyck, M.K., Harding, J.C.S., Fortin, F., Cheng, J., Dekkers, J.C.M. and Plastow, G., 2021. Heritability and genetic correlations of plasma metabolites of pigs with production, resilience and carcass traits under natural polymicrobial disease challenge. Scientific reports, 11(1), pp.1-13. https://doi.org/10.1038/s41598-021-99778-9

Jeon, R.L., C. Gilbert, J. Cheng, A. M. Putz, M. K. Dyck, G. S. Plastow, F. Fortin, PigGen Canada, J. C. M. Dekkers, and J. C. S. Harding. 2021. Proliferation of peripheral blood mononuclear cells from healthy piglets after mitogen stimulation as indicators of disease resilience. J. Anim. Sci. 99(8), p.skab084 https://doi.org/10.1093/jas/skab084

Lim K. S., J. Cheng, A. Putz, Q. Dong, X. Bai, C. K. Tuggle, M. K. Dyck, PigGen Canada, F. Fortin, J. C. S. Harding, G. S. Plastow, and J. C. M. Dekkers. 2021. Quantitative analysis of the blood transcriptome of young healthy pigs and its relationship with subsequent disease resilience. BMC Genomics. 22(1), pp.1-18 https://doi.org/10.1186/s12864-021-07912-8 [7]

Lim, K.S., Cheng, J., Tuggle, C.K., Dyck, M., Canada, P., Fortin, F., Harding, J., Plastow, G. and Dekkers, J.C., 2022. PSXIII-B-14 Investigation of the Blood Transcriptome of Young Healthy Pigs to Identify Genetic Indicators for Disease Resilience. Journal of Animal Science, 100 (Supplement_3), pp.325-325. https://doi.org/10.1093/jas/skac247.593

Putz, A.M., Harding, J.C., Dyck, M.K., Fortin, F., Plastow, G.S., Dekkers, J.C. and Canada, P., 2018. Novel resilience phenotypes using feed intake data from a natural disease challenge model in wean-to-finish pigs. Frontiers in genetics, 9.

Schmied, J.D., Putz, A., Dekkers, J., Dyck, M., Fortin, F., Harding, J., Plastow, G., Canada, P. and Mallard, B.A., 2018. Mortality Rate and Survival of Pigs Classified by Immune Response Phenotype Using the High Immune Response Technology™. World Congress Genetics Applied to Livestock Production, Auckland, New Zealand. Vol. Biology - Disease Resistance 2 Pages 653

Optimizing Feeding and Management Strategy in and Finisher

Dan Bussières

Groupe Ceres

Impact of Net Energy

What should be the optimal level for Net Energy (NE)?

- High or low energy diet
- For 1 % change in NE → FE will change by 1 %
- Predicting the impact of NE on ADG is more difficult
- Pigs likely adjust their feed intake based on the diet NE
- May have an impact if pigs cannot adjust their feed intake
    - Temperature, crowding, feeder space, health…
- The economic impact will differ based on cost of energy and ingredient used

Antibiotic Growth Promoter (AGP)

- Use of AGP and impact on performance
- Ionophore (Salinomycin-Narasin) – Class 4 antibiotic
- Not used in human
- No incidence on antibiotic resistance

- Not very costly and have been proven to improve performance and provide a nice ROI
- Cost 1.60-1.70 $/t (Salinomycin)
- Precautions if using Tiamulin medication!

Feeder Space/Design/Adjustment

- Importance of feeder size/design?
- Recommendation: 12-15 pig/feeder space for wet/dry feeder
- Importance of linear space as well
- We recommend 1.5 to 2.0 inches of feeder space/pig
- Feeder at 12 inches per feeder space are not recommended anymore
- Optimal 14-16 inches/feeder space

Out of Feed Event

Should we be concerned by pig being out of feed?

- **For how long?**
- **Frequency?**
- **Not that much data in the industry to understand the impact**
- **Lot's of factors to consider**
- **Health, crowding, feeder space, pig behaviour etc...**

Feed Particle Size and Texture

- Pellet vs mash
    - ↑ ADG 4-5% average (0.3 à 12.9)

        - ↓ FE 7-8% average (2.5 à 12.4)
- Particle size reduction
    - 100 microns = impact of 1.0-1.5% on FE
    - 1% FE = 0.85 $/pig
    - If we lower particle size from 750 to 450 microns = 2.55 $/pig
- Particle Size
    - PIC meta-analysis
    - In mash feed, linear improvement in FE when lowering particle size
        - 1200-300 microns – 11 % improvement in FE = 1.2 %/100 microns
    - In pelleted feed, improvement until 500 microns about
        - 1000-500 microns – 7 % improvemEnt in FE = 1.4 % /100 microns
    - 1.30 $/100 microns
    - No significant impact on ADG
    - Balance between performance (cost reduction) vs energy need and impact on health (ulcers)

Space Allowance/Density

- Space allowance can be calculated using a k value
- $Weight^{0.67} \times k \times 10.74$ = sq.feet/pig
- Base on the code, here is the minimum sq.feet
- In finisher base on k value of 0.0335

    - 110 kg = 8.30 sq.feet/pig
    - 120 kg = 8.80 sq.feet/pig
    - 130 kg = 9.30 sq.feet/pig
- Does lower sq.feet vs Code will impact performance?
- In finisher we do not reach maximum density when we ship the last pig
- We reach maximum density just when we start to ship pig
- This mean the calculated density should not be base of final end weight, but rather on average weight before first pull/shipping
- What is the optimal density before to start shipping pig?
- Calculated in kg of weight/sq.feet (or lbs/sq.feet)

Feed Withdrawal

- Impact at the slaughter plant on:
    - Meat quality
        - PSE meat du to fast pH drop – too much muscly glycogen
    - Food safety – carcass contamination, condemn
        - Contamination risk at evisceration with feed in the stomach and GI track
- Economics losses for the producer as well if not done properly
    - Feed waste, transportation/condemn losses
- What need to be considered and what are the economics
- 14-24 hours – ideal 16-18 hours
- Include 5-6 hours min prior to loading + 2-3 hours lairage time at the plant + transport time

- Feed withdrawal mean denying feed access not just shutting off the feed line
- Need to maintain full access to water
- Best is to have a proper loading area
- If not, done in the pen where pigs that are not leaving will have out of feed event 1 to 3 time before being shipped – should be able to compensate, but will they alway do?

Marketing and ROI

- How to make a decision about marketing weight strategy?
- Packer grid – weight, premium etc...
- FE at higher weight and feed cost
  - Do we know well the FE at higher weight?
- Market price
- Barn space limitation
- Packer grid also need to be accounted for
- Impact of too light/too heavy pig on index/premium?
- Also, by increasing weight, are we increasing the risk to loose more pig at the end?
- More time in barn and/or more stress on transport
- 0.1 % increase in late finisher dead = 0.20-0.25 $/head
- It can cancel any potential saving
- Increasing weight doesn't mean necessarily adding more time, it may just well be adapting the shipping strategy

Genetic-Sire Line

- The right choice for the sire line used can have a significant impact on performance
- Multiple trials done over time to compare sire line impact on performance and meat characteristic/quality
- Dam line can also have an impact, but doing evaluation of the impact of the dam line on finisher performance is much more complicated

Water Acidification

- Water acidification is well documented and somewhat well accepted for nursery pig, especially in the first 2-3 weeks post-weaning
- What about lowering water pH and impact on finisher pig performance?
- Not a lot of data to support
    - Some data in salmonella control
- Which pH to target?
- Cost vs benefit?
- Potential positive ROI, but still a high investment cost up front
- Need to validate the dose needed to get the right cost
- Dose of acid will vary with water type/quality and type of acid product used
- Need to run pH titration with your own farm water
- Could we use the acid on/off and get the same performance benefit while reducing the cost

Sorting and Mixing Pig

- Used to create group of uniform pig at:
    - Weaning
    - Entry in finisher
    - At shipping
- Does sorting by size has an impact on performance?
- Does mixing impact performance?

Phase Feeding

- What is the optimal # of phase in finisher
- Phase feed vs precision feeding
- Most of the farm will have 3 to 5 phases from 25 kg to market
- Trial with our current 5 phases program vs 3 phases
- Impact on performances
- Interest for mill/feed manufacturing vs number of phases and size of batching at the mill

REAL PEOPLE. REAL WORK. REAL PURPOSE.

A Survey of Best Management Practices of Swine Farms Across Canada

Ken Engele

Manager, Knowledge Transfer
Prairie Swine Centre

The past three years have been ones for the record books. Starting in 2021 widespread dryness seen throughout western Canada, not seen on this level since 2002, resulted in record high feed costs for many producers. Overall 2021 resulted in a crop that was 40-50 percent of average. As any livestock producer knows, feed costs are always the single biggest factor representing 60-70 percent of the total cost of production. Residual effects of these high grain prices carried throughout 2022 and well into 2023. Throughout this period grain prices increased significantly with wheat and barley prices close to doubling the values seen in the fall of 2020, therefore producers where looking for strategies to minimize the impact of high feed prices throughout this time frame.

Inflationary pressures over the past three year, not seen since the 1970s, have increased the financial burden producers have faced over this period. In 2022, for the first time many producers achieved an average hog price that exceeded \$200/ckg for the entire year. While this is a good news story, the challenge was, at the same time cost of production increased up to \$60/ckg for some producers over this same period. Many producers break evens ranged between \$220-

240/ckg. This change in price relationships meant many producers still struggles with positive margins throughout 2022.

A new year brought a set of new challenges. The positive, a drop in feed cost throughout 2023, between $25-30/pig in many cases, provided producers optimism of an improved financial position throughout the year. The negative, an unresponsive hog price resulted in positive margins, for less than expected number of weeks throughout the summer, but largely remained negative for a majority of the year. While feed cost garner most of attention and rightly so, we should not lose focus on those things that we do on-farm on a daily basis to ensure we are optimizing each phase of our farms. Day-to-day activities can be lost in all the noise of everything going on in the barn. Ensuring we are do all the little things right everyday add up to bigger savings than you might think possible?

How are we doing as an industry?

Two projects funded by Swine Innovation Porc and carried out by Prairie Swine Centre (PSC) and Centre de développement du porc du Québec (CDPQ) examined the adoption of best management practices on farms throughout Canada. The first project audited 24 farms across Canada (2018), with a minimum of two in each pork-producing province, consisting of a questionnaire and an on-site visit. The second project consisted of a survey (2023) of pork producers, throughout Canada, of various best management practices implemented in their operations.

Each project had good representation of size and type of operations across the industry. Size of operations ranged from 300 to 6,000 sows, while farrow-to-finish and farrow-to-wean operations represented approximately 80% of respondents, with the balance being wean-to-finish operations. Focus of each project was similar. They focused on best management practices looking at biosecurity/herd health, feed/feeder management, personal protection, water use/management, in addition to each phase of production (breeding, gestation, farrowing, nursery, and grower-finisher). Results from each project were summarized and compared across projects.

Biosecurity Measures

Overall, the industry seems to be doing a good job at implementing various biosecurity measures in their operations. Individual farms need to calculate their individual risk reward ratio when it comes to biosecurity procedures, as not all farms face the same level of risk. Some things to take into consideration is the disease status of your herd or your neighbours herd, proximity to neighbours, pig density in the surrounding area, diseases circulating in the area (PEDv, PRRS), value of your herd (multiplier), and the ability for you to weather some magnitude of a disease break. These things all change over time, as a result should be discussed with your employees, management team, and herd health veterinarian on a regularly scheduled basis to ensure you are ding everything possible to keep your herd healthy as possible.

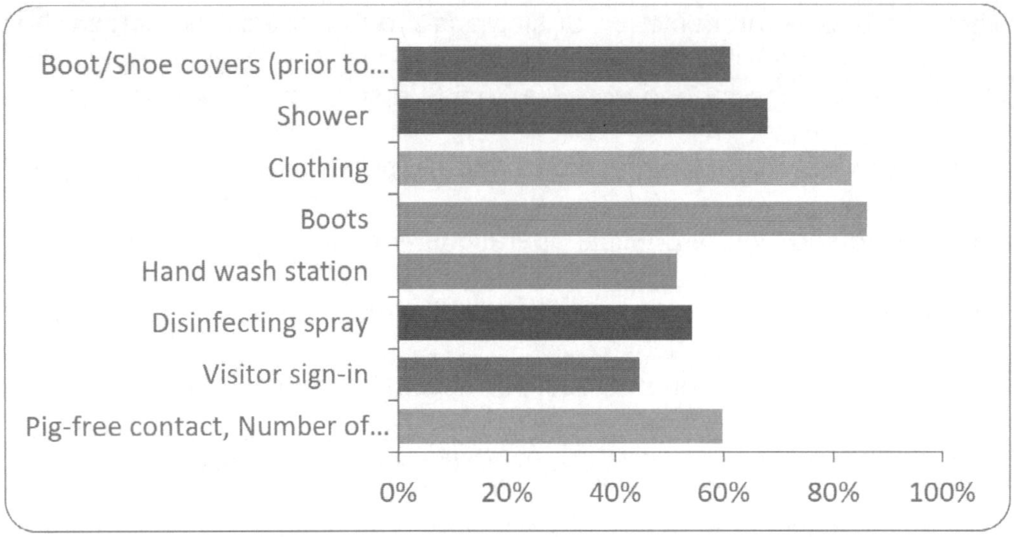

Figure 1. Biosecurity measures implemented on-farm.

Personal Protection

Ensuring a safe work environment is the responsibility of every employer and one that the Canadian pork industry takes seriously. Results in Table 1 indicate pork producers are committed to providing the safest workplace possible for their employees. Audit results indicate that dust masks, hearing protection and hydrogen sulphide (H2S) monitors are being used to varying degrees in on farms across Canada.

While all farms that use H2S monitors use them for pit pulling, it is very important to use them in other key day-to-day activities where H2S could arise. Situations includes power washing and entering the manure transfer station; research indicates that H2S concentrations can exceed acceptable limits any time manure is disturbed. Locations of peak H2S concentrations vary within the room, and vary over time. Therefore, it is essential monitors be provided to all swine barn workers at these key times as H2S may be present in higher than anticipated concentrations. While approximately 60% of participating farms offer H2S training, it is very important that recertification does not get lost in day-to-day activities. Training and standard operating procedures should be provided, at least every three years, so workers can learn how to deal with routine operation and emergencies generating high H2S concentrations.

Table 1. Level of adoption of selected personal protection management strategies.

Category	**Compliant**	**Somewhat compliant**	**Not compliant**
Are dust mask used in the facility? *It is recommended to use dust masks in the facility.*	**83 %**	**0 %**	**17 %**
Is hearing protection used in the facility? *It is recommended to use hearing protection in the barn.*	**100 %**	**0 %**	**0 %**
Are hydrogen sulphide monitors used in the facilities? *It is recommended to always use hydrogen sulphide monitors in the facility.*	**0 %**	**50 %**	**50 %**
Do you provide training on hydrogen sulphide awareness? *It is recommended to provide training regarding hydrogen sulphide awareness.*	**54 %**	**4 %**	**42 %**
Do you provide animal handling training? *It is recommended to provide animal handling training.*	**75 %**	**4 %**	**21 %**

Creep Feeding

Creep feeding is a common practice in swine production, with approximately 90% of farms indicating that they provide creep feed (Figure 2). Most producers generally implement creep feeding 5-7 days prior to the anticipated weaning date.

There are a number of perceived benefits including provision of nutrients, higher weaning weight, and improved transition at weaning, however the benefits of creep feed exist when pigs eat the feed. Intake of creep feed is usually low and highly variable among pigs with approximately 30-50% eating creep feed. It is generally higher in smaller piglets with little to no intake observed in larger piglets. The achieved benefit of creep feeding on growth performance in the suckling and/or nursery period remains inconsistent. Finding new or proven strategies to increase the consumption of creep feed is essential on whether or not utilizing creep feed is important to your operation.

Other work has demonstrated the benefits of providing creep feed is related to enhancing exploratory behaviour in piglets (i.e., allowing natural rooting behaviours) and exposure to feed in a dry form than provision of nutrients. Therefore, it is possible that providing expensive creep diets is not necessary to achieve the benefits of creep feed on weaning weight and overall performance, and that simple diets such as a typical lactation diet, would be sufficient. Creep feed is the most expensive diet used in pig production, costing between $600-1800/tonne. Identification of less expensive alternatives would help to reduce production costs.

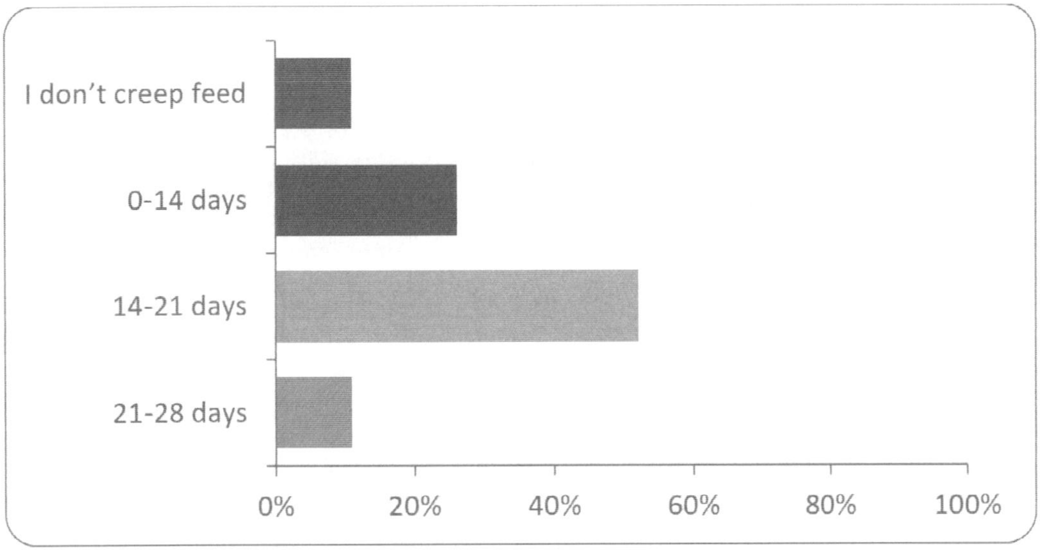

Figure 2. Average age for creep feeding in farms across Canada.

Research has shown that piglets visit tray feeders more frequently compared to the standard feeder. In addition, the different presentation of creep feed appeared to increase the percentage of piglets per litter showing evidence of creep feed consumption. A large tray feeder that encourages social feeding and foraging is more effective at attracting piglets to creep than a standard feeder. The manner, in which we utilize creep feed (to attract piglets), can be improved as a whole (Figure 3) as approximately 40% of respondent use some type of tray feeder.

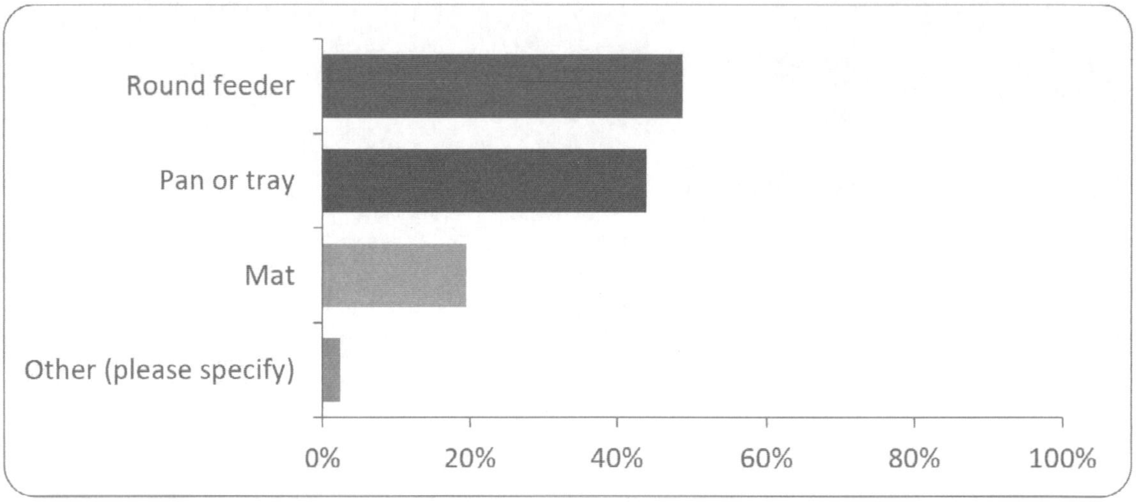

Figure 3. Types and adoption of different types of creep feeders used on farms in Canada.

Results from the survey indicate costs associated with creep feed range from $600-$2,000/mt. However research indicates that overall there appears to be little benefit of providing creep feed in general or of providing complex, expensive creep feed. Specifically work at PSC indicates piglets showed no preference for simple or complex creep feed. In addition, there was little impact of provision of creep feed on pre-weaning performance, with increased ADG only in the final week pre-weaning. While there was a slight benefit to providing creep feed on growth performance in the first week post-weaning, there was no benefit at the end of the nursery.

Enrichment

Enrichment seems to be one of those things that can be easily over looked. The National Farm Animal Care Council's 2014 update to the Canadian Code of Practice for the Care and Handling of Pigs states that all pigs must have "multiple forms of enrichment that aim to improve the welfare of the animals through the enhancement of their physical and social environments." While implementing enrichment on-farm is simple in nature, the proper selection, installation and maintenance of enrichments can have a positive impact to the bottom line of your operation. The lack of enrichment is known to result in more problematic behaviours such as tail-biting and belly-nosing and there is a need for practical and cost-effective solutions that producers can implement.

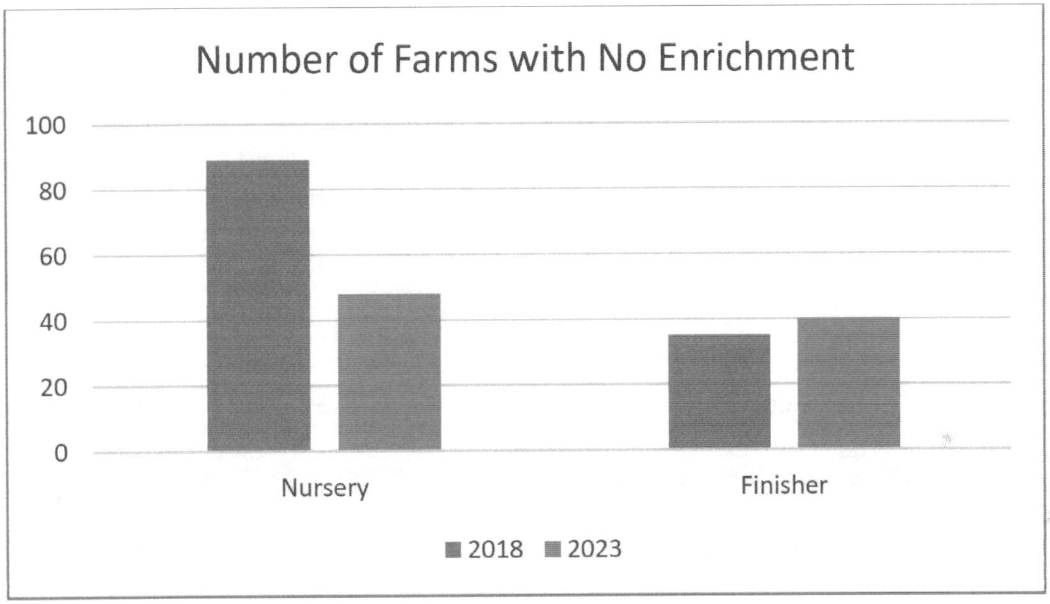

Figure 4. Number of farms with no enrichment 2018-2023.

Based on information presented in Figure 4 enrichment is an area that require additional attention from producers. As seen in Figure 4, just over 50% and 60% of farms incorporate enrichment into nursery and finisher respectively. According to the Code of Practice for the Care and Handling of Pigs (NFACC, 2014) pigs must be provided with multiple forms of enrichment that aim to improve the welfare of the animals through the enhancement of their physical and social environments. The incorporation of enrichment into individual facilities is unique to each operation. Support tools regarding enrichment materials are available in Appendix H of the Code, with different types of enrichment grouped into categories along with the advantages and disadvantages of each type. The Six Ss - Simple Criteria for Choosing Enrichment for Pigs As outlined in Appendix H (page 54) of the Code of Practice there are six criteria to consider when choosing enrichment for your operation. Considerations could include the follows aspects: SIMPLE, SAFE, SANITARY, SITE, SOFT, & SUSPENDED. Figure 2 (below) provides more detail for each of these considerations.

Sorting Pigs

The majority of grower-finisher systems are based on ad libitum feed and water intake. Although access to feed or water may be temporarily limited by another pig, all pigs are able to obtain as much of these resources as they want. As the number of pigs per feeder space increases, pigs adapt their eating behaviour to reduce the effect of this restriction. Pigs increase their rate of eating (gram/min), and decrease their total duration of eating (min/day) when feeder-stocking density (pigs/feeder space) increases. As long as they are able to obtain adequate amounts of feed, dominant pigs respond by eating more quickly rather than increasing their defence of the feeder.

Reducing the floor space allowance of pigs is known to reduce feed intake and average daily gain. When housed in crowded conditions, there is a loss in productivity for at least part of the feeding period. The impact of crowding, results in depressed production across all social groups, and is not affected by dominance. Research conducted at Prairie Swine Centre studied the impact of sorting pigs as they enter the grower-finisher phase.

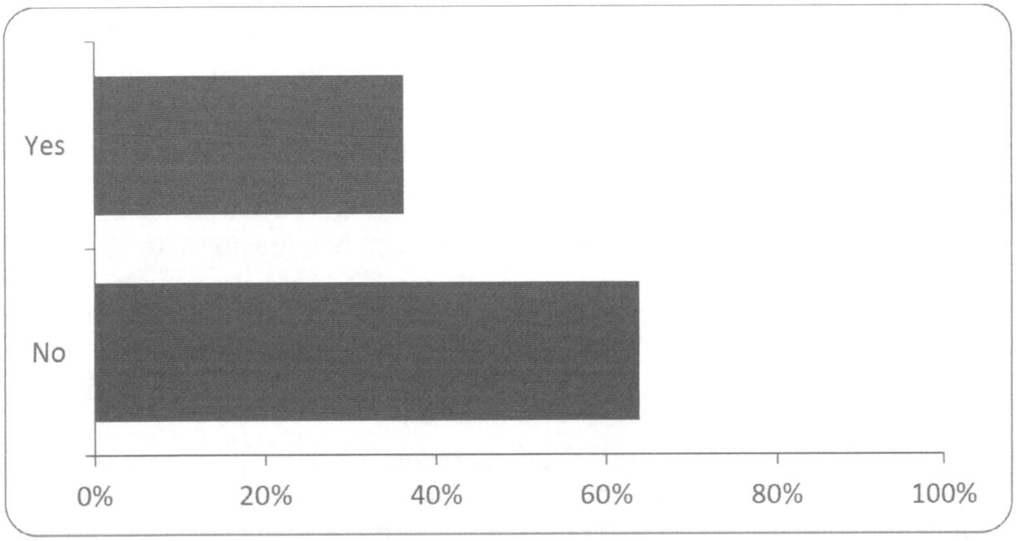

Figure 5. Percent of pigs sorted by size entering the finishing phase

Results from the audit indicate that approximately half of participating farms sort pigs when transferred into the finishing barn. There are multiple factors to consider when deciding to sort or not at this phase. Research indicates that finisher pigs fed ad libitum do not benefit from sorting. However, sorting on the basis of nutritional needs can be effective in cases such as split-sex feeding, as well as for newly-weaned pigs, in order to provide the best diets to each group.

Water Use and Management

Among nutrients, water is required in the greatest amount but quite often receives the least attention. Water intake of finisher pigs has been reported to range up to three times feed intake, depending on body weight and feed intake. However, most 'water intake' reported is in the form of water disappearance from drinkers, including water wastage, rather than water actually consumed by pigs. Previous work has shown finishing pigs can waste 25% of water from well-managed nipple drinkers, therefore opportunities exist to reduce wastage when flow rates are adjusted on a regular basis1.

Actual on-farm water flow rates and nipple drinker heights were measured on 24 farms across Canada, representing each phase of production from gestation to finishing. Note that not all farms had nipple drinkers installed in each phase of production, for example, some producers solely relied on wet/dry feeders without an additional water source. Table 2 outlines water flow parameters showing ranges measured for low, target, high, and very high values. Recommended flow rates should range between 1.0 to 2.0 L/min and 0.5 to 1.0 L/min for farrowing and all other phases of production respectively, while the target range used in the analysis was expanded from 0.5 to 1.5 L/min for all areas other than farrowing.

Table 2. Water flow recommendations

	Low (l/min)	**Target (l/min)**	**High (l/min)**	**Very High (l/min)**
Gilt pen	<0.5	0.5-1.5	1.5-2.5	>2.5
Gestation	<0.5	0.5-1.5	1.5-2.5	>2.5
Farrowing	<1.0	1.0-2.0	2.0-3.0	>3.0
Nursery	<0.5	0.5-1.5	1.5-2.5	>2.5
Finishing	<0.5	0.5-1.5	1.5-2.5	>2.5

Table 3. Measured water flow rates of selected farms across Canada (%)

	Low (l/min)	Target (l/min)	High (l/min)	Very High (l/min)
Gilt pen	5.1	33.33	56.4	5.1
Gestation	0	59.4	21.9	18.8
Farrowing	15.3	38.9	29.3	16.6
Nursery	15.2	56.8	19.0	8.9
Finishing	5.4	29.3	54.3	10.9

Overall water management within audited farms varies across phase of production (Table 3). Generally, producers do a better job in managing flow rates within Gestation (pens) and Nursery, where approximately 60% of the nipple drinkers measured met the target flow rate. The challenge is in Finishing, where approximately two-thirds of nipple drinkers provide flow rates in excess of pig's requirement, with 11% of nipple drinkers being rated very high (>2.5 L/min).

Conclusion

Based on some of the results we can see that little changes can make a big impact on the overall profitability of your operation. While most producers are aware of individual best management practices throughout their barns, day-to-day activities and emergencies sometimes get in the way. Currently there seems to be a margin for improvement as approximately 40% of measured and surveyed best management practices are achieved. I don't think doing 100% of things 100% of the time is possible. If we can move that needle incrementally from 40 to 50% then to 60% can save producers substantial dollars over the long run.

BACK TO BASICS
START PIGS RIGHT
Getting pigs started right is more important than ever with today's challenges. With a proven track record, AP 920 plasma can help you get the job done quickly, easily and effectively. Sometimes getting back to the basics is the best solution.
100% Bovine | 100% Porcine | UV Processed
APCproteins.com
WATCH THEM thrive
APC

The Effect of Space Allowance and Marketing Strategy on Production Performance of Grower-Finisher Pigs

Willem Steyn* and Jaco Raath

SwiNE Nutrition Management, Pretoria, South Africa.
wjsteyn@swine.co.za

1. Introduction

In modern pig production, optimizing stocking density is a critical factor for both animal welfare and economic sustainability. A majority of the fundamental research regarding space allowances for grower-finisher pigs was conducted several years ago with older genotypes and much lighter marketing weights than today's standards. Higher number of piglets born alive and increasing market weight trends over the past years, reflect the evolution of the swine industry. These trends are driven by a combination of factors, including genetics, nutrition, management practices, and market demands. Producers are motivated to push these trends to further dilute fixed facility cost.

This study delves into the multifaceted relationship between stocking density and marketing strategy, assessing their combined impact from the grower phase to slaughter. With an increasing global demand for high-quality pork products, understanding how stocking density and different marketing strategies influences pig performance, carcass quality and economics, is of paramount importance.

This research not only sheds light on the welfare and productivity of pigs under varying stocking conditions but also provides valuable insights of potential strategies on how we can produce more meat per square meter space, utilizing and optimizing our financial return on the capital invested. While market conditions are difficult for pig producers to control, producers can control their feeding strategies or delivery moments.

2. Definitions

Space allowance refers to the number of square meters (m^2) of floor area provided per pig in a grower - finisher production facility. Sufficient space is essential for pigs to exhibit natural behaviors, reduce stress, and optimize growth. Inadequate space can lead to increased aggression, reduced feed intake, and compromised pig welfare. To express floor space allowance, k-value was used, as it describes pen space as a function of effective pig space requirements (body weight $^{0.67}$). The k-values were calculated using a formula reported by Whittemore (1998): space per pig (m^2) = k x BW $(kg)^{0.67}$. According to Gonyou et al. (2006), a k-value of 0.0336 is required for maximal feed intake and ADG for grow-finish pigs with fully slatted flooring. Table 1 below indicates the number of pigs that can be kept per 1000 m^2 barn floorspace. This was derived using the target weight development curve of the Topigs Norsvin Tempo progeny.

Table 1. The number of pigs per 1000 m^2 barn space, based on their live weight (kg).

Age (days)	Weight (kg)	Space/Pig (m^2/Pig)	Number of Pigs (Nr)
70	29	0.32	3118
77	35	0.36	2749
84	41	0.4	2472
91	47	0.44	2256
98	54	0.49	2056
105	62	0.53	1874
112	69	0.57	1744
119	77	0.62	1621
126	85	0.66	1517
133	93	0.7	1428
140	101	0.74	1351
147	109	0.78	1284
154	117	0.82	1225
161	124	0.85	1178
168	132	0.89	1129
175	140	0.92	1086

Marketing strategy involves the pre-selecting of individual pigs before marketing, also referred to as topping. It has become a common strategy to pre-select the heavier pigs in a batch, making room for the others to fully grow out. This is done to come as close as possible to homogenous batches and reduce the variation within a batch. The non-selected pigs are normally the inefficient slow growers, taking up valuable space. The additional space the pigs are given for the last week(s) will theoretically allow them to recover growth and reach higher weights at market. Ultimately, topping strategies are used to maximize facility space while minimizing reduced performance from high pen stocking rates.

3. Objectives

The objective of the trial is to evaluate the response of grower-finisher pigs to a range of different scenarios based on stocking density and marketing strategy by looking at the growth and performance of pigs during the grower and finisher phases as well as measuring the necessary carcass characteristics at slaughter and determining the economics of production.

To develop and test practical concepts to better utilize existing grower unit facility space to produce more meat per square meter, keeping within the recommend space allowance per pig to allow pigs to exhibit natural behaviors and optimal growth. With the current societal concerns regarding animal welfare, pig producers must maximize production efficiency and minimize facility cost, while at the same time, meeting animal welfare practices.

4. Trial Design

The trial was conducted at the Baynesfield SwiNE Research Unit at Baynesfield Estate in KwaZulu Natal, South Africa. A total of 528 animals (31.7 ± 0.15 kg initial BW, Topigs TN 70 x Tempo genetics, 70-days of age) were used in an 84-day study, consisting of the grower-finisher period. The house consists of 48 pens each containing 10 to 20 pigs with a total of 528 pigs in the house. Pens were randomly allocated to one of six treatments (n = 8 replicas), with treatment one acting as the control. All treatments were fed the same diet from the start of the grower phase up to slaughter consisting of four feeding phases. Ractopamine was included at 5 ppm in the last feeding phase. Individual body weights and pen feed intakes were taken weekly for treatments one to three. Body weights and pen feed intakes were collected at every phase change, up to the last phase, where it was collected weekly until marketing for treatments four to six. Pigs had *ad libitum* access to feed and water. Additionally, the carcass parameters lean meat percentage, carcass weight, and dressing percentage were measured at slaughter. Feed cost, carcass value, margin over feed cost (MOFC), and cost per gain were calculated. Statistical analyses were conducted in SAS (2021) utilizing the Proc GLM model for the average effects within each period and connection letter calculated with Fisher's test.

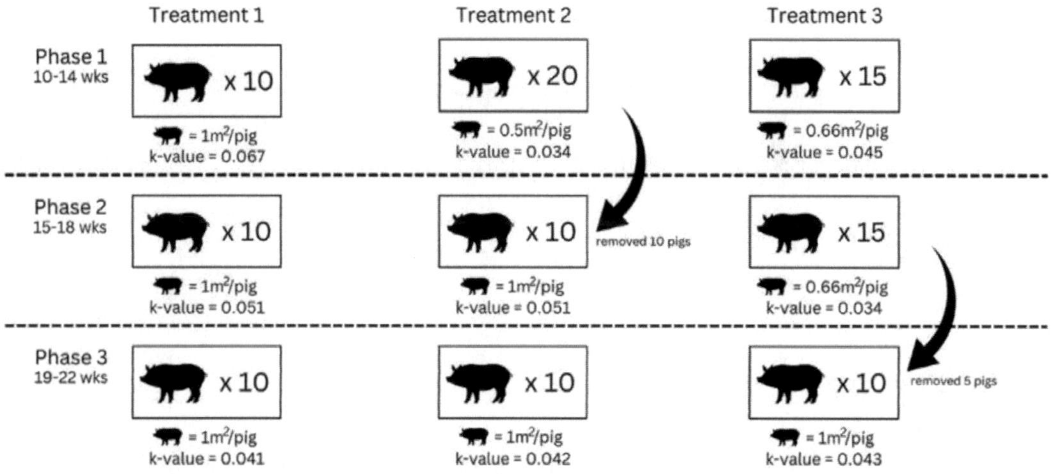

Figure 1. Schematic illustration of the experimental design for Treatment 1,2 and 3 (Space allowance)

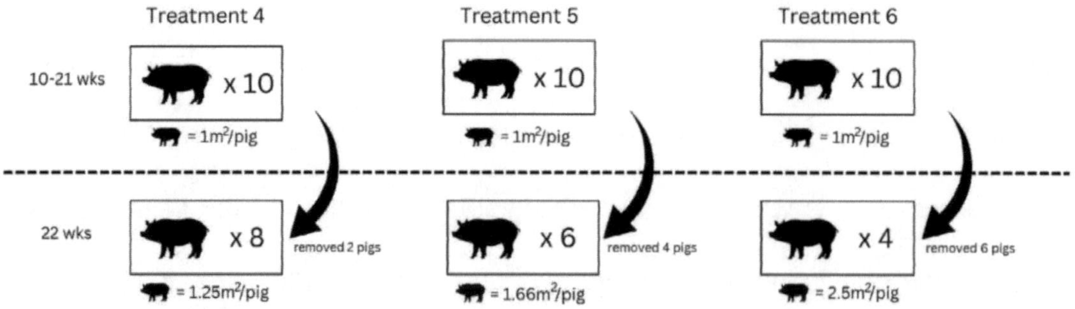

Figure 2. Schematic illustration of the experimental design for Treatment 4,5 and 6 (Marketing strategy)

5. Results

The first part of the trial focused on the space allowance during the grower-finisher phase.

The results provide insights into the overall performance of the trial. Based on the results in Table 2 it is evident that Treatment 1 had significantly better individual production results compared to Treatment 2 and Treatment 3. Despite the removal of animals at the 14 and 18-week marks, there was no recovery growth observed to offset the loss in production.

Table 2. The effect of different stocking densities on overall performance for Treatment 1, 2 and 3 (Space allowance).

	Units	Treatment 1	Treatment 2	Treatment 3	SE
Start Weight (Kg)	Kg	31.67	31.74	31.75	0.83
End Weight (Kg)	Kg	115.62	113.27	110.92	1.81
Per 1 ADG	Kg	0.885^{a}	0.831^{b}	0.862^{ab}	0.02
Per 2 ADG	Kg	0.994^{a}	0.963^{ab}	0.948^{b}	0.02
Per 3 ADG	Kg	1.036^{a}	1.007^{ab}	0.978^{b}	0.01
AFI Per 1	Kg	47.33^{a}	39.26^{b}	41.23^{b}	0.98
AFI Per 2	Kg	64.55^{a}	65.81^{a}	61.36^{b}	0.7
AFI Per 3	Kg	78.43	78.34	76.93	1.48
Total AFI	Kg	190.32^{a}	183.41^{b}	179.52^{b}	2.52
FCR Per 1		1.94^{a}	1.63^{b}	1.67^{b}	0.03
FCR Per 2		2.23^{a}	2.34^{b}	2.28^{a}	0.03
FCR Per 3		2.62^{a}	2.74^{b}	2.99^{c}	0.08
Total FCR		2.26^{a}	2.24^{a}	2.31^{b}	0.03

$^{a, b,}$ Means with different superscript are different ($P < 0.05$).

Based on these results, we simulated scenarios to illustrate the practical application of the data, presented in Table 3. Each treatment's results were expressed in relation to 1000 m^2 space availability. This approach further illustrates the influence of space allowance on the economic viability of rearing pigs under different scenarios.

Table 3. The economics based on 1000 m^2 space allowance for Treatment 1, 2 and 3.

	Units	Treatment 1	Treatment 2	Treatment 3
Total Available Space	**m^2**	1000	1000	1000
Total Cost of Space	**CAD**	250000	250000	250000
Animals Sold	**Number**	1000	1000	1500
Cold Carcass Weight	**Kg**	88.1	93.2*	79.8#
Age at Market	**Days**	154	161*	145#
Pork Price/kg	**CAD**	2.6	2.6	2.6
Income/Pig	**CAD**	229.1	242.2	207.5
Total Income	**CAD**	229112	242206	311220
Feed Cost/Pig	**CAD**	80	84.8*	76.1
MOFC /Pig	**CAD**	149.1	157.4	131.4
Total Feed Cost	**CAD**	80000	84805	114075
Total MOFC/Batch	**CAD**	149112	157401	197145
MOFC/Area ($/$m^2$)	**CAD**	149.1	157.4	197.2

*Due to double stocking pigs during the first phase, it opens the possibility to slaughter pigs one week older (later). The assumption was made based on gain during the last weeks, that pigs will put on an additional 5 kg dressed weight. Additional feed consumed in that last week have been calculated on the same basis.

#Based on the trial design, one third of the pigs went to market at 18 weeks of age to correct for the space, while the rest of the pigs continued till 22 weeks of age. Weights and feed intake due to early marketing were considered.

The second part of the trial focused on marketing strategy and topping pigs one week before marketing. No significant differences were observed in the overall production results considering the different topping strategies portrayed in Table 4.

Table 4. The effect of removing animals on performance for Treatment 4, 5 and 6 (Marketing Strategy).

	Units	Treatment 1	Treatment 4	Treatment 5	Treatment 6	SE
Average Start Weight (kg)	**Kg**	31.67	31.58	31.74	31.57	0.77
Average Slaughter Weight (kg)	**Kg**	115.62	116.02	116.90	116.06	1.55
Overall Gain (kg)	**Kg**	83.95	84.45	85.16	84.49	1.03
Overall ADG (kg)	**Kg**	1.024	1.030	1.039	1.030	0.01
Overall FI (kg)	**Kg**	2.31	2.40	2.39	2.37	0.03
Overall FCR		2.27	2.33	2.31	2.31	0.03
Dressing Percentage (%)	%	79.69^{a}	78.71^{b}	78.71^{b}	78.41^{b}	0.21
Cold Carcass Weight (kg)	**Kg**	88.12	87.27	87.97	86.90	1.23
Lean Meat Percentage (%)	%	71.34	71.07	70.91	71.33	0.12

$^{a, b,}$ Means with different superscript are different ($P < 0.05$).

Results from 21 to 22 weeks of age indicated a significant positive reaction on individual performance by topping of animals one week before marketing. Animals in Treatments 4, 5 and 6 respond positively to the additional space in the last week before market. Graphs 1,2 and 3 below illustrates the average daily gain, feed intake and FCR between the different treatments.

Graph 1. The effect of removing pigs on average daily gain during the last week of the finisher phase

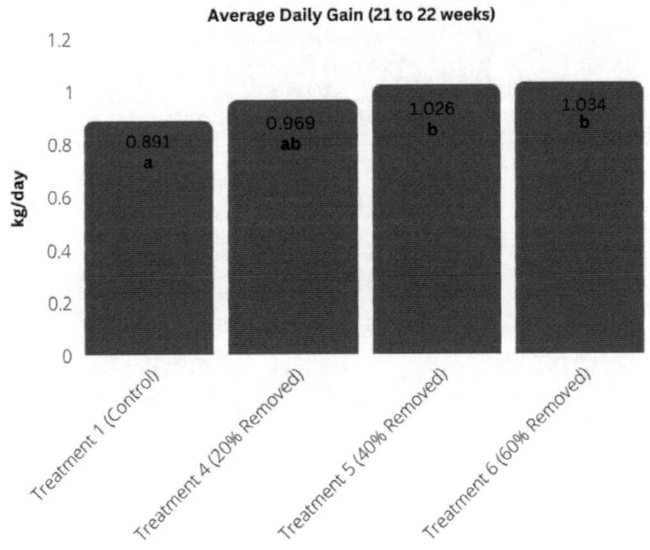

Graph 2. The effect of removing pigs on average daily feed intake during the last week of the finisher phase.

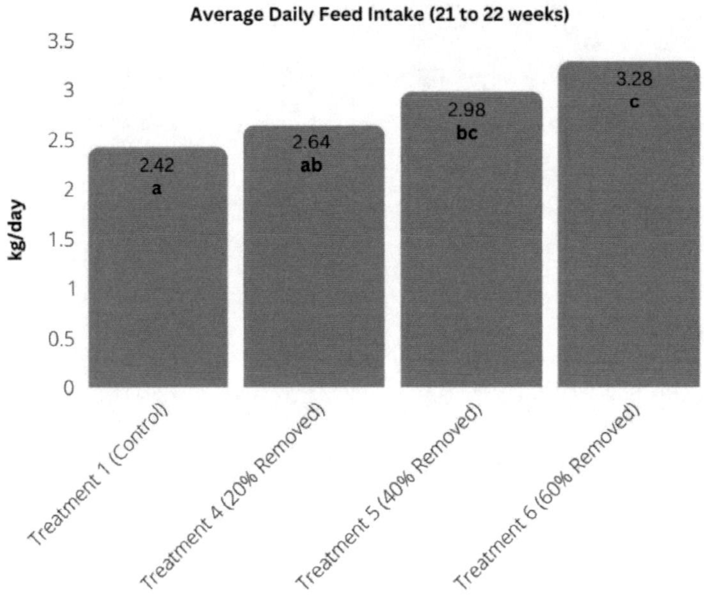

Graph 3. The effect of removing pigs on feed conversion ratio during the last week of the finisher phase.

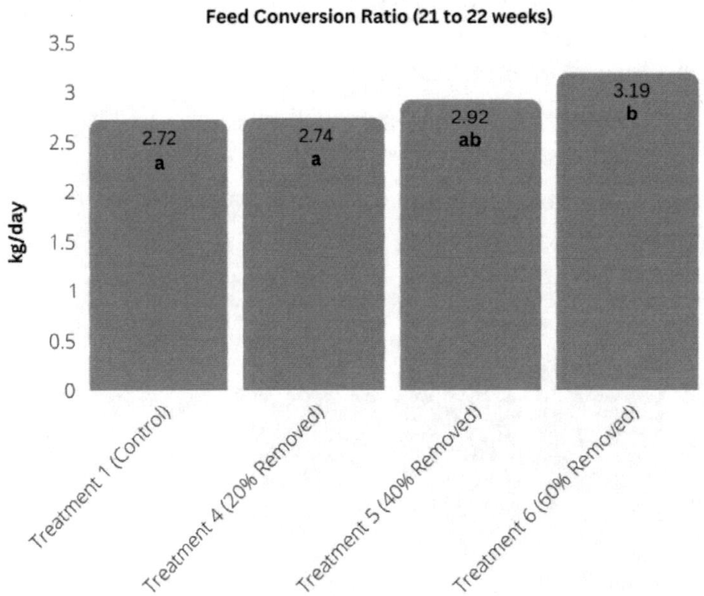

6. Conclusion

The standard approach to pork production requires a mind shift from focusing solely on individual animal performance to optimizing meat production per square meter of available barn space. This approach not only maximizes the total meat yield and optimizes economic return but also ensures efficient utilization of space availability.

Strategies such as adjusting space allowances through stocking density and/or marketing strategy can play a pivotal role in achieving this goal. By optimizing space allocation, we can accommodate more grower-finisher pigs in the same area without compromising their welfare. This not only increases the meat yield per square meter but also improves the overall productivity of the farm.

In essence, the future of pig farming lies in strategic space management and a holistic approach that focuses on total meat production. By adopting these strategies, we can ensure a sustainable and profitable future for the pig farming industry. It's time we look beyond individual performance and embrace a more comprehensive and efficient approach to meat production.

References

Gonyou, H. W., M. C. Brumm, E. Bush, J. Deen, S. A. Edwards, T. Fangman, J. J. McGlone, M. Meunier-Salaun, R. B. Morrison, H. Spoolder, et al. 2006. Application of broken-line analysis to assess floor space requirements of nursery and grower-finisher pigs expressed on an allometric basis. J. Anim. Sci. 84:229–235. doi:10.2527/2006.841229x

Lerner, et al., 2020. Effects of space allowance and marketing strategy on growth performance of pigs raised to 165 kg. Anim. Sci. 2020.4:1252–1262 doi: 10.1093/tas/txaa065

O'Quinn P.R., Dritz S.S., Goodband R.D., Tokach M.D., Swanson J.C., Nelssen J.L., Musser R.E. Sorting growing-finishing pigs by weight fails to improve growth performance or weight variation. J. Swine Health Prod. 2001;9:11–16.

SAS Institute Inc. (2021). SAS/STAT® 16.1 User's Guide. Cary, NC: SAS Institute Inc.Gerbing, D. (2014).

Whittemore, C. T. 1998. The science and practice of pig production. 2nd ed. Malden (MA): Blackwell Science Ltd.

South West Vets
BETTER TOGETHER
SWINE HEALTH
PRODUCTION SUPPORT
APPLIED RESEARCH
INFORMATION SERVICES
CANADIAN PORK EXCELLENCE VALIDATIONS
At South West Vets, we work together to enhance your business - from health to production management, and everything in between.
(519) 271-2111
www.southwestvets.ca

Fostera Gold PCV
Fostera Gold PCV MH
THREATS ADAPT
SO SHOULD YOUR PROTECTION
BROADEST ANTIGENIC AND
LONGEST LASTING PCV2 PROTECTION
zoetis

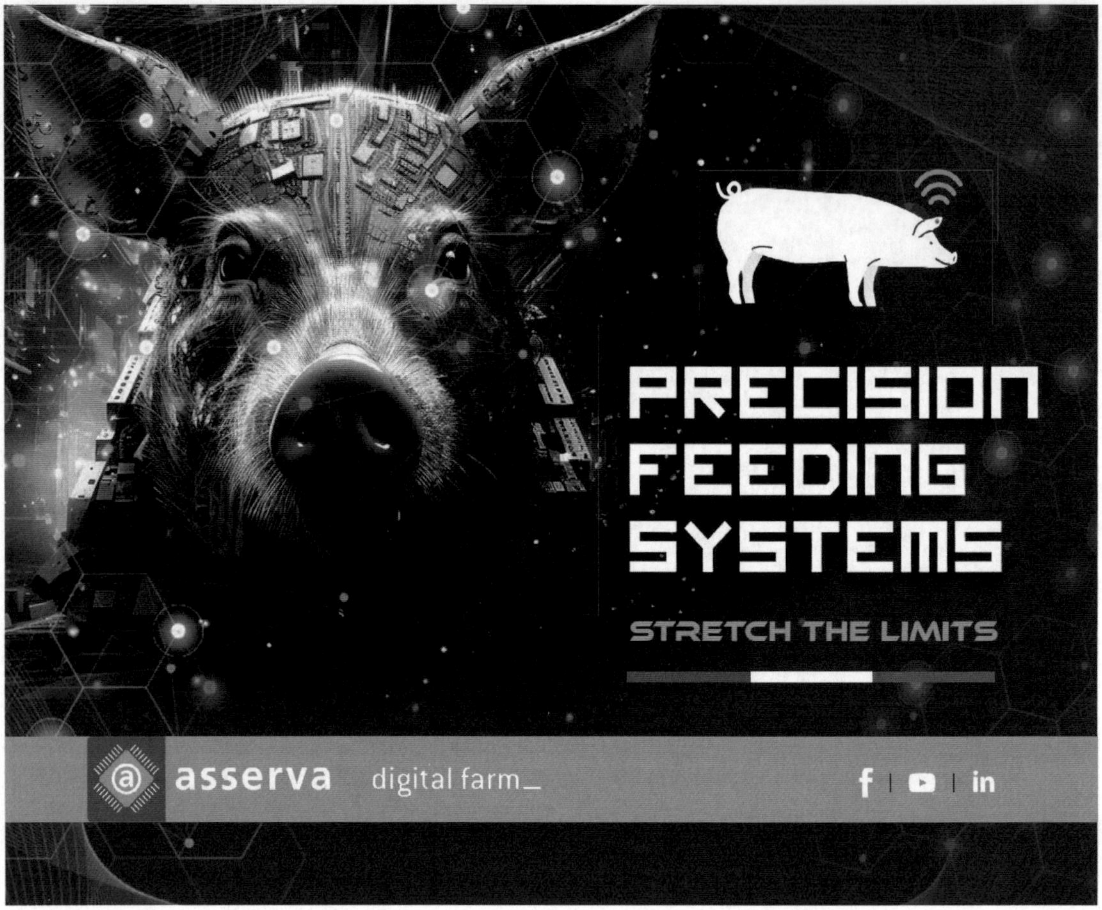
PRECISION
FEEDING
SYSTEMS
STRETCH THE LIMITS
asserva
digital farm_
f | in

Reliable Inputs,
Sustainable
Outlook
watson
SWINE PERFORMANCE INVESTIGATOR
CO_2
$
www.trouwnutrition.ca
trouw nutrition
a Nutreco company

Where Have All the Gilts Gone, Impact of Mortality & Removals on Gilt Retention

Jayne Jackson

PigCHAMP, Inc., 1531 Airport Rd., Ames, Iowa, 50010
Jayne.Jackson@PigCHAMP.com

Gilt retention or sow longevity continues to be a widely discussed topic not only for the economic impact it has but the welfare concern to our industry, it is one of the industries hot buttons where we struggle to have the right combination of actions for improvement. There are so many different factors that can impact and influence sow retention from genetics, nutrition, health, environment to management practices. With all these factors it is hard to narrow down where to start and what actions will make the largest impact on improving the retention within your own organization.

Using the PigCHAMP North American database representing over 2,100,000 gilts. This analysis will spotlight when, where and why gilts and sows are being removed from the herd in the hopes to help guide you in a direction to narrow down areas within your own production to focus. The database was created using the PigCHAMP report Gilt Retention Analysis customized to include days since serviced, removal reasons, if euthanized, and other production day intervals. This data set represents multiple genetic suppliers, various farm types and sizes, management practices, etc. with gilt arrived between 2015 – 2020; with removals and subsequent performance into August 2023. For this analysis we will be looking primarily at parity zero through parity four.

Retention Rate

The definition of "Retention Rate" it is the percentage of gilts introduced in a period that achieve consecutive farrowings. Over the past four years we have seen a drop in retention between sows first parity and their second. In both 2019 and 2020 roughly 83% of gilts arrived in those years farrowed their first parity of those same sows only 65-66% continued to have their second parity.

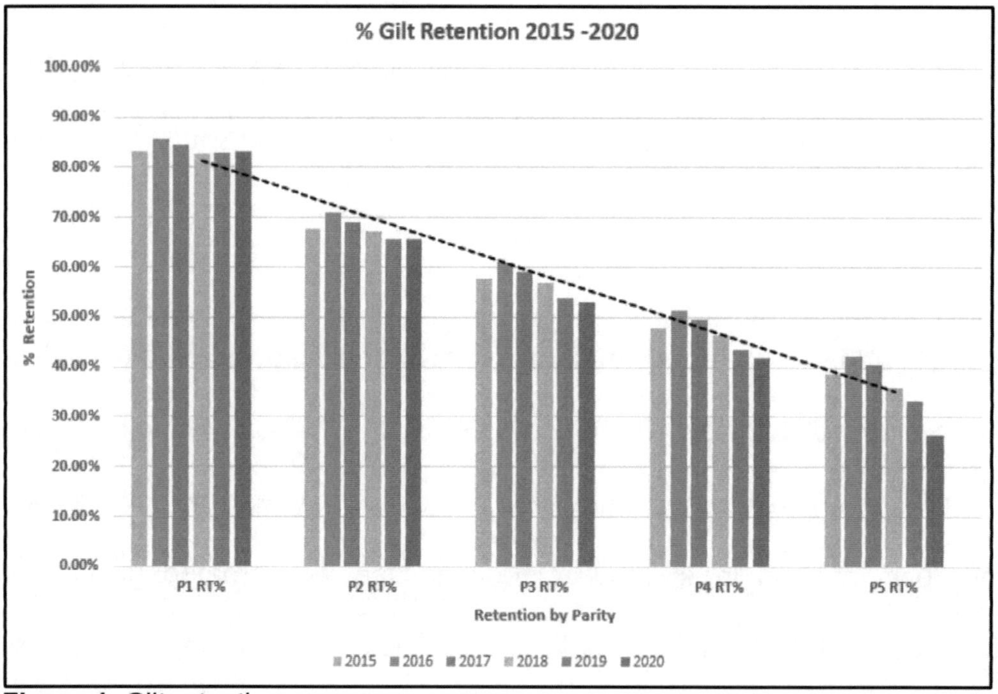

Figure 1. Gilt retention

Data suggests the highest productivity is associated with parity three sows and many studies show a sow will not return a positive value net income until reaching parity three or four (Stalder et. al., 2003; Gruhot et. al, 2017) but as an average the industry is only retaining roughly 48% of the gilts to a third litter. Taking that in account let's look at retention a bit differently so instead as the precent of sows that did continue farrowing let's look at it as a loss between the parties. As you can see the time between first parity and second parity is an area that can and should be focused on.

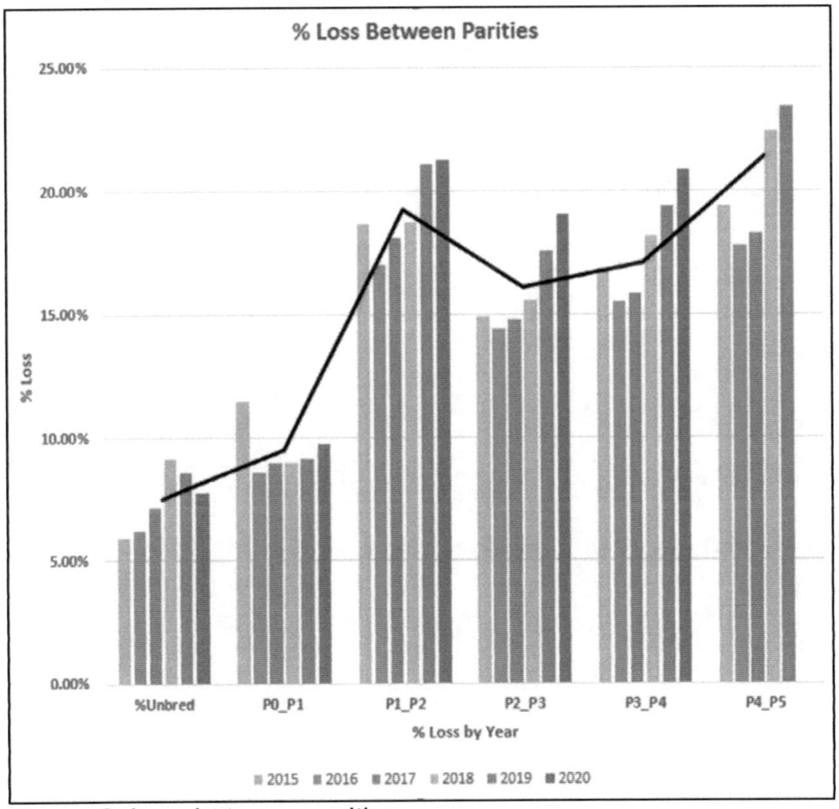

Figure 2. Loss between parities

To enable us to look at where, when, and why the gilts and sows are falling out earlier in life we included the reason removed in the data set, we excluded any females/sows removed do to research, herd audits (clean records), depopulation, transfers & natural disasters. We have also grouped the multitude of reasons to basic categories and for analysis P0 represents bred gilts, unless otherwise stated.

Sow Morality

When discussing retention sow mortality has the most focus at this time primarily due to issues related to prolapse. There has been extensive investment and research looking into the cause, affect, and potential solutions for prolapse with in the industry as an example, researchers are currently working at various methods that can be used to identify sows with a greater risk for pelvic organ prolapse (POP) incidence. A team of researchers from Iowa State University have developed and tested a three-point perineal scoring system based on visual evaluation of the perineal region of late gestation sows (Kiefer,Studer,Ross 2023).

Looking at where the mortality occurs over 50% of sow's deaths occur in the farrowing barns of that 33% of the deaths happen between day 112 thru 122 a 10-day window of time, much of that contributed to prolapse however I think there is some additional opportunities for improvement related to mortality.

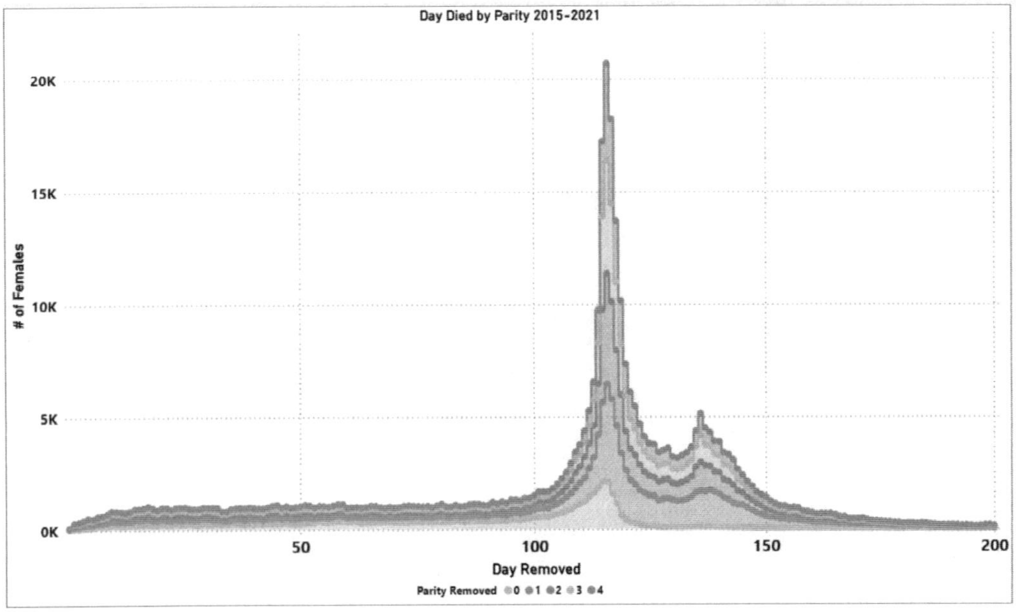

Figure 3. Day removed

When breaking down the “why” between day 110 -125 you will see the prolapse very prominently standing out in red however there are other areas with opportunity that can have an impact especially related to earlier parities, such as lameness, farrowing complications, sudden death and other/unknown. Prior to delving into these types of deaths it is recommended that there are clear definitions and understanding of the reasons for the farm and management teams. In the case of farrowing complications and especially related to sudden death and other/unknown you may want to expand on these generalizations.

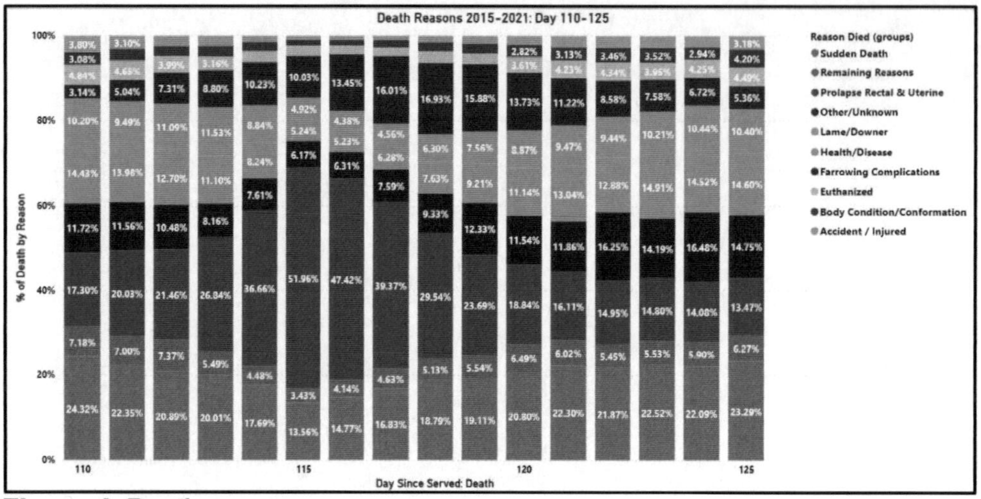

Figure 4. Death reasons

Cull Rate

Though sow morality has been fore front I would argue that larger focus should be related to the cull rates in earlier parities. When early removals occur due to reproductive performance the key issue can be associated with the management of the gilts from birth to the entry into the herd (Patterson and Foxtrot 2019). As an industry we are struggling to get these young gilts and sows to their second parity and beyond to their most optimal production.

We can see where these removals are happening over a 200-day period based on days since served prior to cull. As expected, we see fallouts between day 21-60 during gestation, again around farrowing time with the majority of removals starting to occur on day 140 -170 after weaning coming back into heat and subsequent gestation cycle.

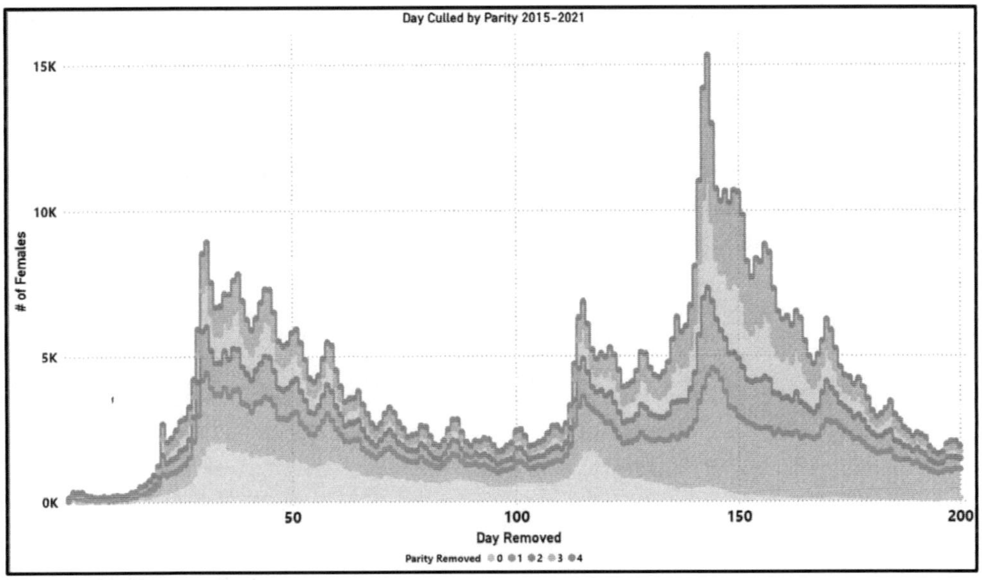

Figure 5. Day culled

When focusing on the reasons for removal over the same period for parities zero - one nearly 60% of the culls are related to reproductive failure. This holds true up until the 140–170-day period where reproductive failure is nearly 39% of the loss, locomotion (lame) 16% with farrowing performance at 10.5%.

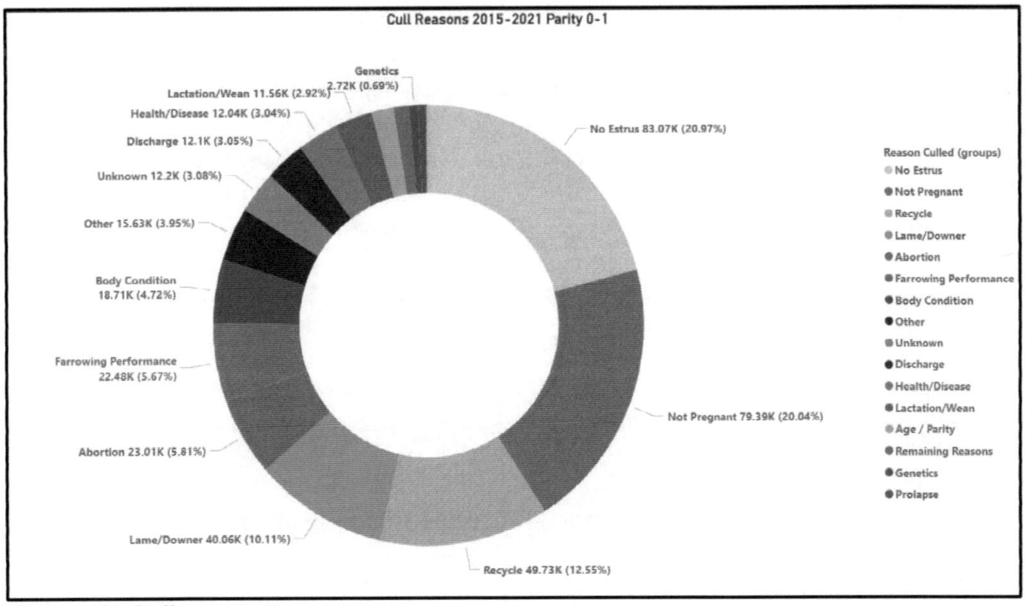

Figure 6. Cull reasons

Building a strong foundation is important to herd retention and production, with most losses related to reproductive performance and in the early stages of the female's life the focus should be on selecting the right gilts and managing those gilts through their early development into productivity. The successful introduction and retention of gilts through their early parities drives lifetime performance of the breeding herd and represents an opportunity to improve and enhance overall sow longevity (Patterson, Canavate 2023).

Conclusion

Understanding the farms retention is important to determining where to start in identifying areas that can be influenced and impacted by changing processes and/or adopting new practices.

I suggest using the days since service as a marker for removals to help identify the best areas to initially focus. The advantage of this number is the ability to see the full picture including those sows that were subsequently bred back and then were removed. If using the removal reasons to narrow the focus further, I caution on the number of reasons and lack of definition. Consolidate into primary groups and validate the farm has a clear understanding when providing the reasons for analysis.

With a better understanding of sow mortality related to prolapse and there are researchers involved in finding ways to identify those at-risk females and looking for various solutions it is time for us to move our focus to what I feel the larger issue is, the loss of those earlier parity females.

The most impact on retention related to the early parities would be to react to the reproductive failures. Here basic practices such as identifying the correct gilts/sows are in heat, identifying prolific staff members with good AI practices to train additional staff, creating processes that are gilt focused minded, etc. will make an impact on the overall performance and production of the herd.

Implementing gilt development programs (GDU) where knowledgeable staff can solely focus on the selection and proper management of the gilt for success on the farm. Though with labor shortage it is difficult to see devoting the time here but in the long run I think it will have the greatest return.

Bottom-line, the suggestions above are not anything necessarily new some may say they are basic husbandry skills we just need to get back to practicing on the farm, validating they are implemented correctly and if not training accordingly.

References

Gruhot, T., J.A. Calderon-Diaz, T. J. Baas. et al. 2017. An economic analysis of sow retention in a United States breed-to-wean system. J. Swine Health Prod. 25(5):238-246.

Kiefer,Z.E.,Studer, J.M,Ross, 2023. Improving pig livability by understanding pelvic organ prolapse https://www.pigchamp.com/news/benchmark-magazine/articles/ArtMID/2128/ArticleID/306/Improving-Pig-Livability-by-Understanding-Pelvic-Organ-Prolapse

Patterson, J., Foxcroft, G. (2019). Gilt Management for Fertility and Longevity. Animals, 9(7), 434. https://doi.org/10.3390/ani9070434

Patterson,J., Canavate,S., 2023 Sow Retention and Productivity https://www.pigchamp.com/news/benchmark-magazine/articles/ArtMID/2128/ArticleID/305/Sow-Retention-and-Productivity

Stadler K.J., C. Lacy, T.L. Cross, et al. 2003. Financial impact of average parity of culled females in a breed to-wean swine operation using replacement gilt net present value analysis. J. Swine Health Prod. 11(2):69-74.

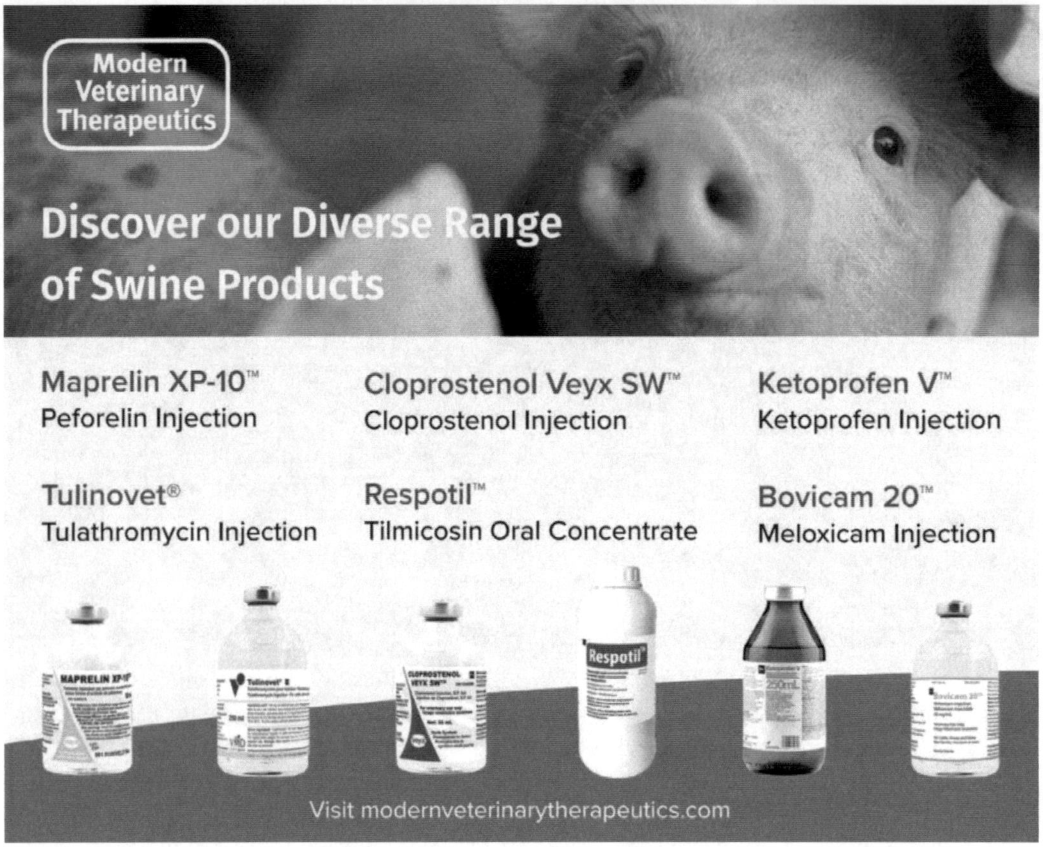
Modern Veterinary Therapeutics
Discover our Diverse Range of Swine Products
Maprelin XP-10™
Peforelin Injection
Cloprostenol Veyx SW™
Cloprostenol Injection
Ketoprofen V™
Ketoprofen Injection
Tulinovet®
Tulathromycin Injection
Respotil™
Tilmicosin Oral Concentrate
Bovicam 20™
Meloxicam Injection
Respotil
Visit modernveterinarytherapeutics.com

Maximize your reproductive success with Minitube.
minitube
www.minitube.com
More piglets thanks to high quality products and hygiene concept for insemination
800-500-7978
info.canada@minitube.com

Proposition 12 Sow Management: Lessons Learned

P.J. Corns

JBS Live Pork, LLC, Greeley, CO

Summary

Conversion of farms from pen gestation to California Proposition 12 compliant housing takes on a multitude of different dynamics. From overall building structure and farm flow considerations to labor allocations, the entire production process has required an exhaustive review. A multitude of factors continuously impact normal pen gestation, albeit we have well over 30 years of experiences to draw from of which we know the causes associated with each effect. The adoption of Proposition 12 compliant production simply has willing participants swimming in uncharted waters through a multitude of new and existing retrofits.

Having sound and practical production practices in areas such as body condition scoring and training of gilts pre-breeding to free access stalls (FAS), are always critical to the overall performance of sow farms. Having them in place and executed to the highest degree is even that more important to the success of a Prop.12 conversion. The ability to analysis the impact of moving sows pre- and post-implementation has had a significant impact on building new build strategies. One example of this would be the layout of the pen after moving sows, such as determining the proper length of paneling used to create stanchion penning. As with any new endeavor without experience, our forecasted calculations for extra labor proved to be very close to reality in our system. Lastly, we have learned that identifying and segregating at risk animals has proven challenging. This is not only due to implementation, but other various factors such as building capacities and the interpretation of this by the California Department of Agriculture, from both a health and welfare standpoint.

Within the JBS Live Pork system, conversion from open pen gestation to California Proposition 12 compliant housing has tested our production acumen, construction concepts, and overall stockman ship on the farms. The ability to adapt, compartmentalize, and execute on our standard operating procedures has taken this animal housing option from theory to reality and into a space that we feel we are growing better at. The ability to stay fluid has better positioned JBS to adapt to a landscape that as farmers, we inherently have little sway over; politics namely the constituents of California and the Supreme Court of the United States. By doing so, we keep our eye on the future of pork production as it is dictated ultimately by our end customer needs.

THE BOAR WITH
BRAGGING RIGHTS.
Working with a producer-owned, 70,000 full-program sow herd, we are the only swine genetics company with direct access to data from commercial production through meat packing. The industry-leading standard — imitated but never duplicated — is our DNA Line 600 Duroc.
DNA
GENETICS

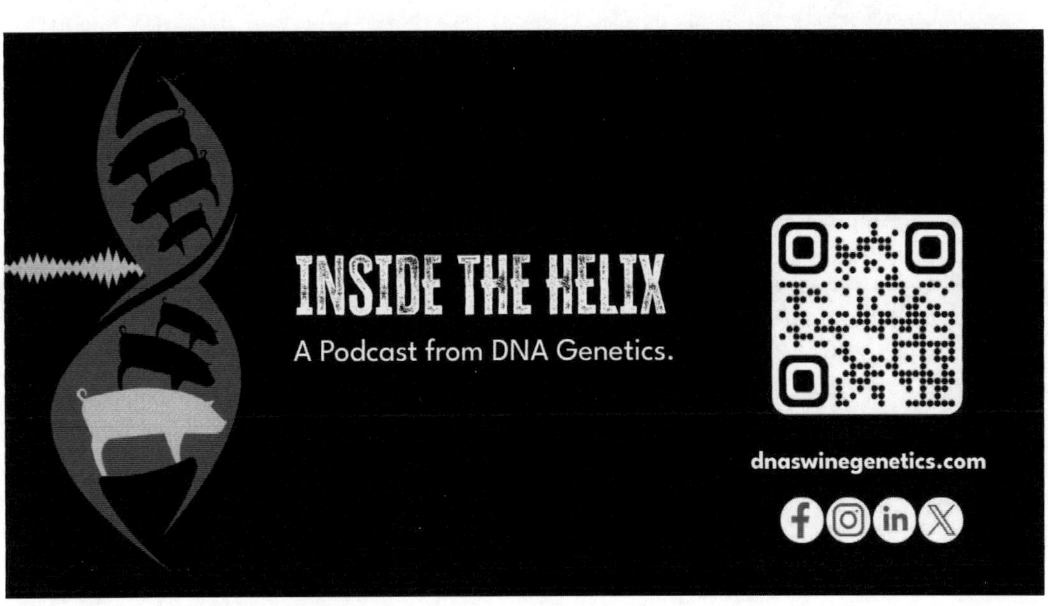
INSIDE THE HELIX
A Podcast from DNA Genetics.
dnaswinegenetics.com

Internal Biosecurity – Is There Value for You?

Brad Chappell & Mikala Sasley

Topigs Norsvin Canada Inc.; 20 South Landing Dr. Unit #1, Oak Bluff, MB R4G 0C4
brad.chappell@topigsnorsvin.ca

There are 3 types of biosecurity that are commonly discussed in swine production. Namely external, biocontainment and internal. Our industry has done an excellent job discussing external biosecurity, but less emphasis has been placed on biocontainment and internal biosecurity. A definition of each type would be:

- External Biosecurity – prevention of pathogens (diseases) from entering your barn.
  - You do not want a PRRS break.
- Biocontainment - prevention of pathogens (diseases) from exiting your barn.
  - If you have a PRRS break you want to keep it on your farm and NOT share with your neighbors
- Internal Biosecurity - prevention of pathogens (diseases) from spreading within your barn.
  - If you are trying to eliminate PRRS you do not want to keep cycling the virus in your barn

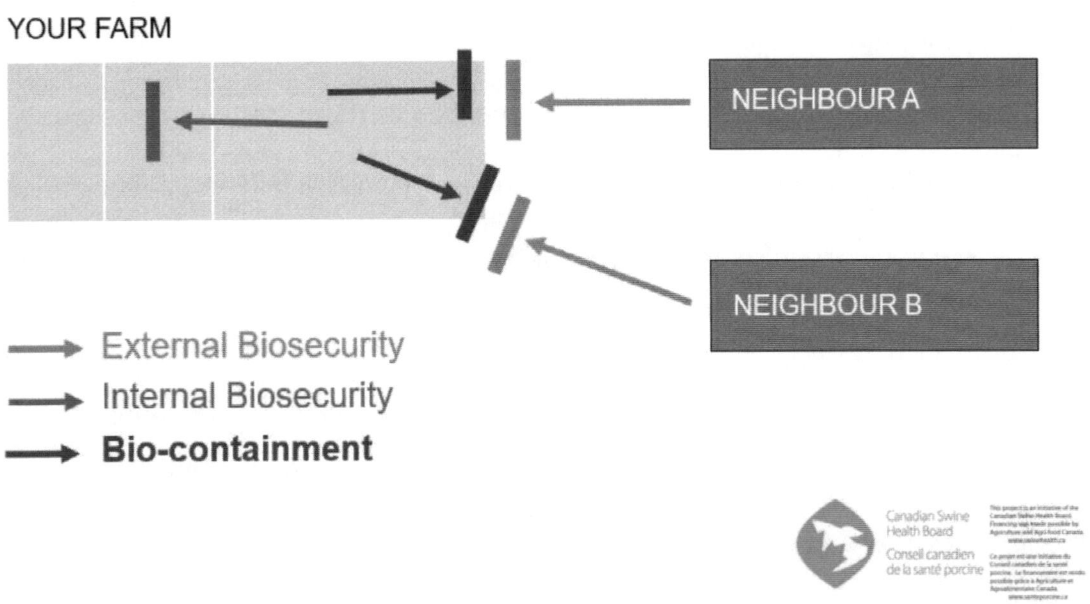

Figure 1. Three Types of Biosecurity

My presentation focuses only on internal biosecurity and the benefits of practicing this type of biosecurity on your farm daily. If you have had to deal with PED elimination, then you will have a good understanding of internal biosecurity and biocontainment during the PED elimination process.

Several examples of internal biosecurity that could be routinely completed:

1. Manure Scrapers:
    1.1. Do you have separate scrapers for each room? If not, what a great way to spread piglet scours between farrowing rooms! Even ileitis in the finisher!
2. Syringe Sanitation:

2.1. If your syringes are not kept clean or ideally sterile you are injecting your pigs with bacteria! Is this causing swollen joints or antibiotic resistance on your farm?

3. Do you wash your boots between each section or room?

   3.1. The same theory applies here – you do not want to take the post weaning rotavirus scour from the second oldest nursery room back to the youngest nursery room!

4. Do you wear gloves while doing chores?

   4.1. Not wearing gloves as you do chores and changing gloves between rooms will help to spread bacteria or viruses on your hands.

5. Do you routinely wash and disinfect your hallways?

   5.1. Significantly helps to reduce the disease load in the barn!

6. Water Lines & Water Cisterns & Water Medication Tanks.

   6.1. Do you keep them clean?

   6.2. Dirty water lines of cisterns are a sure way to spread nursery scours due to E coli.

7. Piglet Processing Carts:

   7.1. What is your sanitation program for the cart? What a great way to spread rotavirus piglet diarrhea.

   7.2. How do you keep your processing tools clean?

8. Dead Stock Management:

   8.1. Do you wash, disinfect, and dry the pails or carts used to take dead stock to the disposal room? If not, what pathogens are you cycling between the rooms?

Update on Sapovirus: An Industry Perspective

Matthew A Ackerman, DVM[1]; Jennifer Demare, DVM[2]; Dylan L Lape, DVM[1]

[1]PorkVet Solutions, 6110 W US Hwy 52 New Palestine, Indiana 46163 USA
Matt.Ackerman@PorkVet.com
[2]PremierSHP, Steinbach, Manitoba, Canada

1. Herd History

The story begins with a phone call I received from a gilt customer of one of my clients. My client had been a closed herd gilt supplier since 2016. They had been able to maintain themselves PRRS naïve and PED naïve the entire time they were in multiplication. They were selling isowean gilt pigs to a customer in Iowa.

Despite my client's squeaky clean health status, the customer and Iowa said that pigs on arrival had a rotavirus A scour that was leading to an increase in mortality and was not responding to medication. At the time my client had no clinical signs of anything at the sow farm.

About three weeks later, my client started having scouring in the sow unit in piglets at 7 to 10 days of age that appeared like Coccidia with the pasty yellow mucoid diarrhea. Unfortunately, the diarrhea did not respond Marquis® paste (15% w/w ponazuril Boehringer Ingelheim Animal Health) and instead of just having low-level morbidity we started having mortality taking our pigs wean per sow from a 11.5 to a 10.0 and going from weaning 1,200 pigs a week to 1,000 pigs a week in the 2400 sow unit.

2. Response

Our initial response was to perform Oral Live Virus Exposure (OLVE – aka feedback) which appeared to work 4 to 6 weeks later when the sows & gilts exposed pre-farrowing made it into the farrowing house. However, after the initial bump the OLVE really didn't seem to improve health nor production and pigs wean per sow continued to fall.

We then implemented 4 week and 1 week pre-farrow vaccination with the Sequivity RP vaccine (Merck) with a Sapovirus & Rotavirus C in addition to our standard pre-farrow vaccination program. Pre-weaning mortality seem to stabilize, and pigs weaned per sow improved as well. It wasn't an immediate complete recovery; however, over the next six months, health and production returned to normal.

As we continued our Sequivity vaccine with the Sapovirus and Rotavirus C in addition to our regular vaccine pre-farrow scour vaccine (E coli, Clostridium C & Rotavirus A), herd health stabilized and the downstream nursery complaints resolved as well.

3. Cost of the Disease

If you look at the cost of the disease in my client's herd (Figure 1), Sapovirus took their annual wean pig production from 65,000 pigs weaned (on an average year) to 58,355 pigs weaned in this 12 month period or cost them 6,645 pigs weaned that year. If you take the average wean pig value as $54 CA ($40 US), then Sapovirus cost my client $358,830 CA ($265,800 US) in that 12-month period on this 2400 sow farm.

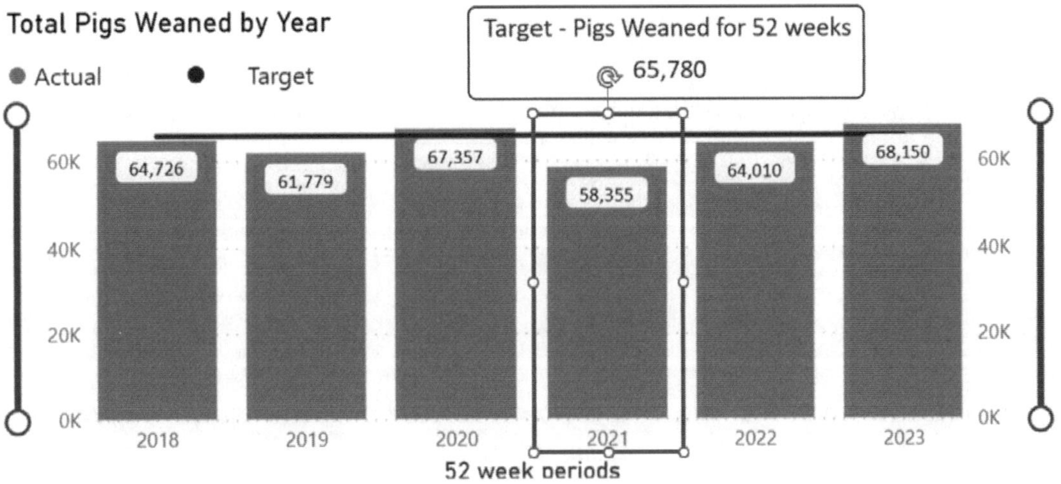

Figure 1. Pigs weaned by year.

This translates to about $150 CA per sow ($111 US) or about $65 CA per sow farrowed ($48 US).

4. Return On Investment (ROI)

From the ROI standpoint, this vaccine (Rotavirus & Sapovirus) would cost about $3 Canadian per dose times two doses or $6 per sow. So, at $65 CA per sow farrowed / $6 CA cost per sow of vaccine = 10 to 1 Return on Investment if you herd would have experienced this level of morality.

5. What Had Others Seen?

Others are seeing similar reporting issues, whether that experiencing a reduction pigs wean per sow due to increased morbidity or reduced wean weight. (SHIC 2023) Will Fombelle, DVM reported that they lost a pound per pig of wean weight, which he reported cost them about $29 CA ($22 US) per sow per year.

$29 CA per sow per year / $13.80 CA (cost of 2 doses x 2.30 litters per year) = 2.15 to 1 ROI.

6. Conclusion

From my experience in the field, it is my professional opinion that Sapovirus:

a. Is a real pathogen
b. Causes scours at 5 to 10 days of life
c. Can affect nursery pigs as well
d. In affected herds, Replicon Particle (mRNA) vaccines are well worth pursuing

7. Acknowledgments

We would like to thank the Banff Pork Seminar for the opportunity to share our experiences with you. I also appreciate the practitioners and producers that allowed me to share their information in this presentation.

References

Swine Health Information Center (SHIC) – (2023) September 2023 SHIC eNewsletter – Swine Health Information Center https://www.swinehealth.org/september-2023-shic-enewsletter/

Swine Health Information Center (SHIC) – (2015) SHIC Porcine Sapovirus – Fact Sheet - 24Apr2021.pdf (swinehealth.org) https://www.swinehealth.org/wp-content/uploads/2021/04/SHIC-Porcine-Sapovirus-Fact-Sheet-24Apr2021.pdf

AQINAC
Association québécoise des industries de nutrition animale et céréalière
Les Éleveurs de porcs du Québec
présentent l'événement du secteur porcin
present the pork industry event
LE PORC SHOW
CONFÉRENCES • EXPO • FESTIVITÉS
CONFERENCES • EXHIBITION • CELEBRATIONS
CENTRE DES CONGRÈS DE QUÉBEC
QUEBEC CITY CONVENTION CENTRE
DÉCEMBRE 2024 DECEMBRE
DE RETOUR POUR UNE 11e ÉDITION
BACK FOR A 11th EDITION
Pour rester à l'affût de l'information concernant Le Porc Show, suivez notre page Facebook.
To stay up to date with information regarding Le Porc Show, follow our Facebook page.

ABSTRACTS: Best of Current Research

Nutrition

Health

Reproduction Management

Behaviour & Welfare

No difference in growth performance feeding frost-damaged or mid-tannin, hulled or dehulled faba bean to weanling pigs

L.F. Wang[1], J.A. Valle[2], F. González[2], P.W. Nyende, R.T. Zijlstra[1] and E. Beltranena[1,3]

[1]University of Alberta, Edmonton, AB, Canada T6G 2P5; [2]University Autónoma de Baja California, Mexicali, México 21705; [3]*Email*: eduardo.beltranena@ualberta.ca

Nitrous oxide (N_2O) is ~300 times worse than carbon dioxide (CO_2) in global warming potential. The main agricultural offender is the application of nitrogen (N) fertilizers that may volatize or leach into water bodies if not managed properly. Rhizobia that form nodules in the root of faba bean plants make it the legume that fixates the most atmospheric N and thus requires no N fertilizers. Faba bean provides 35-40% starch and 25-30% protein displacing cereals and imported soybean meal in feeds. However, the faba bean crop takes several weeks longer to grow than field pea making it more susceptible to frost damage. Colour sorting grey to blackened hull beans prioritizes nondamaged beans for human consumption. It was not clear if frost-damaged or mid-tannin faba bean should be fed to hogs. To test these hypotheses, 240 weaned pigs were fed a) nondamaged Snowbird [6 g/kg tannins], b) frost-damaged Snowbird, c) dehulled nondamaged Snowbird), d) dehulled frost-damaged Snowbird, e) mid-tannin Florent [14 g/kg tannins] or f) dehulled Florent faba bean over 2 growth phases (14 d each; 20 or 30% inclusion) starting 1 wk post-weaning. Feeding any dietary treatments resulted in NO effect on weekly or overall growth performance or final trial body weight (Table 1).

Table 1. Growth performance of weaned pigs fed frost-damaged or mid-tannin, hulled or dehulled faba bean.

Variable	Snowbird hulled	Snowbird dehulled	Frost-damaged hulled	Frost-damaged dehulled	Florent hulled	Florent dehulled	SEM
Feed disappearance, g/d	876	844	854	875	880	846	22
Weight gain, g/d	586	572	566	592	599	583	12
Gain-to-feed, g/g	0.67	0.68	0.67	0.68	0.68	0.69	0.008
Final BW, kg	25.25	24.84	24.82	25.58	25.41	25.07	0.44

Implications: Feeding frost-damaged Snowbird or mid-tannin Florent faba bean to weanling pigs resulted in NO difference in growth performance. Dehulling, to removed the frost-damaged blackened hull or tannins concentrated on the hull, had no benefit. Pork producers may therefore source nearly mature, grey to blackened, frost-damaged faba bean at reduced cost to lower feed cost. Producers must also cash carbon credits including legumes in feed formulations.

Effects of source of ingredient, diet particle size, and sex on growth performance and gastric ulcer incidence in finishing pigs

B.M.H. Chau, A.D. Beaulieu and R.W. Newkirk

Department of Animal and Poultry Science,
University of Saskatchewan, Saskatoon, SK, Canada

Air-classified (AC) pea starch is a promising ingredient for pork producers, however the very small particle size may negatively impact pig health and performance. The objective of this experiment was to determine if the diet particle size from AC pea starch inclusion affects gastric ulcer incidence and growth performance. A 2 × 2 × 2 factorial experiment (2 ingredient sources (pea vs. corn) × 2 particle sizes (fine starch vs. ground grain) × 2 sexes), with 176 pigs (initial body weight (BW) of 82.1 ± 3.8 kg) were allotted to 1 of 4 diets in a completely randomized block design with 12 replicates per treatment for 35 days (3 blocks, 4 pigs per pen). Pelleted diets were formulated to contain 0 or 20% AC pea starch or corn starch, resulting in fine (500 um) or coarse (800 um) average particle size diets. Weekly individual BW and feed intake were obtained to calculate average daily gain (ADG), average daily feed intake (ADFI) and gain-to-feed ratio (G:F). One hundred and twenty-six stomachs, obtained at slaughter were visually scored to assess the treatment effect on stomach ulcers according to a 5-point scoring scale (0. normal to 4. severe). No differences in BW, ADG, ADFI, or G:F were observed due to AC pea starch inclusion ($P > 0.10$). Barrows had a higher ADFI than gilts (2.76 and 2.56 respectively; $P \leq 0.01$). The overall G:F improved with decreased diet particle size regardless of ingredient (0.45 and 0.43 for 500 and 800 µm respectively; $P \leq 0.01$). An interaction in G:F between sex and particle size was observed but only when gilts were fed a finely ground diet (0.48 vs. 0.42, 0.43 and 0.44 respectively; $P = 0.01$, sex by particle size). No interactions were observed for BW, ADG, ADFI, or G:F between sex, source of ingredients or particle size ($P > 0.10$). Most pigs had ulcers but there was no effect on ulcer scores due to diet particle size or ingredient ($P = 0.23$ and 0.14 respectively).

Implications: Finely ground starch, including AC pea starch, can be used in diets at 20% inclusion without affecting the severity of gastric ulcers or growth performance of finishing pigs however the reduction in particle size improved G:F.

Effects of beta-glucans on the intestinal inflammation in weanling pigs

Marley Conceição dos Santos[1,2,3], Josiane Carla Panisson[2], Renata Bacila[3], Vivian Vieira[3], Alex Maiorka[3] and Simone Gisele de Oliveira[3]

[1]Department of Animal and Poultry Science, University of Saskatchewan, Saskatoon, SK, S7N 5A8; [2]Prairie Swine Centre, Inc., SK, S7H 5N9; [3]Department of Animal Science, Federal University of Paraná, Curitiba, PR, 80035-000
Email: marley.santos@usask.ca

The objective of this study was to evaluate the effect of feeding diets with high levels of soybean meal and the inclusion of a yeast-based beta-glucan additive, on the performance and immunity of weanling pigs. A total of 96 mixed-sex nursery piglets (2 piglets/pen) with an initial body weight of 5.9 ± 0.8 kg were randomly assigned to 1 of 4 dietary treatments (n=10 pens/treatment) in a completely randomized block design for a 42 d feeding trial. Dietary treatments were a complex diet (CD; standard diet with a greater inclusion of highly digestible ingredients) and simple diet (SD; with more than 30% soybean meal), with or without the addition of the yeast-based beta-glucan additive. Pigs were fed ad libitum and weighed weekly. Blood samples were taken on day 7 and 21. Intestinal permeability, cytokines and phagocytosis profiles were measured from the blood collected. No significant differences were reported in piglet performance between any of the variables analyzed ($P > 0.05$). The CD with the additive showed the lowest intensity of granulocyte phagocytosis and IFN-α concentration at 7 days ($P < 0.05$). SD down-regulated the concentration of IL-6 cytokines and up-regulated IL-8 ($P < 0.05$). The addition of yeast-based beta-glucan additive to the diet showed a down-regulation in IL-8 and the intensity of monocyte phagocytosis was observed at 7 days, as well as an increase in the intensity of monocyte phagocytosis at 21 days, and an increase in the concentration of phagocytic granulocytes at 7 days ($P < 0.05$). These results show that a diet with a higher concentration of soybean meal may increase inflammation in the gastrointestinal tract of nursery pigs, while the addition of the yeast-based beta-glucan may reduce this inflammatory response. Furthermore, the interaction between a CD and the beta-glucan-based additive also reduces intestinal inflammation.

Implications: Beta-glucans and a complex diet may reduce damage to the animals' gastrointestinal tract.

Multi-NSP enzyme activities improve performances of piglets consuming high fiber and low net energy diets

Leni Kuterna[1], Bertrand Medina[3*], Dan Bussières[2] and Ivan D. Girard[3]

[1]Palital Feed Additives B.V., Gelderland, 5334 LH Velddriel, Netherland; [2]Groupe CÉRÈS Inc., St-Nicolas, Québec, G7A 3S8, Canada; [3]Probiotech International Inc., St-Hyacinthe, Québec, J2S 8L2, Canada
*Email: bertrandmedina@probiotech.com

Both soluble and insoluble non-starch polysaccharides (NSP) may limit intestinal absorption and digestion of dietary nutrients along the intestinal tract of swine. AveMIX®XG$_{10}$, a commercially available enzyme-based feed additive with high endo-1,4-β-xylanase and endo-1,3(4)-β-glucanase activity was evaluated on nursery pigs consuming high fiber and low net energy (NE, -56 kcal/kg feed) diets. A 47-day trial was conducted with 21-day old nursery pigs (6.0 kg) allocated to three groups: control (CON; 2456 kcal/kg feed), low energy (LE; 2400 kcal/kg feed), and low energy + enzyme (LE+E; 2400 kcal/kg feed + AveMIX®XG$_{10}$ at 100 g/MT on of feed). All groups had 13 pens of 28 pigs (n=364), and received wheat and corn mix based diets that had a similar NSP content (11.4%). For each pen the daily feed intake, weekly weight, diarrhea index and mortality were recorded. Data was analyzed by an ANOVA model (Statistix 8.0; Analytical Software, FL) including treatment and pen as fixed and random effects, respectively, and a Chi-square test for pairwise differences. Compared to both CON and LE, LE+E increased (P <0.05) average daily gain (ADG) and average daily feed intake (ADFI) during weeks 4-7 and weeks 1-7, and significantly increased final weight. During weeks 1-3, ADG and ADFI did not differ between CON and LE+E groups, and the LE+E group was significantly higher than LE. Over the entire period, LE impacted (P <0.05) negatively the feed conversion ratio (FCR) compared to CON, however, during weeks 4-7 the LE+E had similar FCR compared to CON (P>0.05).

Implications: An enzyme-based feed additive with high endo-1,4-β-xylanase and endo-1,3(4)-β-glucanase was shown toimprove piglet performance when dietary NE content is decreased which may offer some formulation flexibility and profitability.

Effect of feeding barley-based diets differing in fermentable starch and fibre on nutrient digestion and body composition in weaned pigs

J. Sanchez-Zannatta[1], L.F. Wang[1], E. Beltranena[1], A.D. Beattie[2], R.W. Newkirk[2] and R.T. Zijlstra[1,3]

[1]University of Alberta, Edmonton, AB, Canada T6G 2P5; [2]University of Saskatchewan, Saskatoon, SK, Canada S7N 5A8; [3]*Email*: ruurd.zijlstra@ualberta.ca.

Feeding barley grain to weanling pigs is not common in western Canada. However, both hulled (woody fibre, rapid digestible starch) and hull-less (fermentable fibre, slow digestible starch) barley offers the opportunity to investigate how young pigs could benefit from its dietary inclusion. The range in starch and fibre content among barley cultivars presents opportunities to investigate whether distinctive characteristics of barley grain influence nutrient digestion, fermentability (microbial gut production of short chain fatty acid [SCFA]) and body composition. To explore, 40 pigs (8.6 kg; 1-week post-weaning) were fed 1 of 4 diets containing 600−705 g of cereal grain/kg: 1) low-fermentable wheat (LFW); 2) high β-glucan, hull-less barley (HFBB); 3) high amylose, hull-less barley (HFAB); 4) low-fermentable hulled barley (LFB). Diets provided 10.2 and 9.8 MJ net energy (NE)/kg, and 1.32 and 1.21 g standardised ileal digestible (SID) lysine/MJ NE for phase 1 (day 0–14) and phase 2 (day 15–28), respectively. Apparent ileal digestibility (AID) of β-glucan was 3 times greater for LFW than HFBB and HFAB, whereas AID of starch was 12%-units lower for HFAB than other diets. The AID of crude protein (CP) and gross energy (GE) did not differ among diets. Apparent total tract digestibility of GE and CP was 4.2 and 8.9%-units lower, respectively, for the 3 barley diets than the wheat diet. The apparent hindgut fermentation (AHF) of β-glucan was 2.6 times greater for HFAB and HFBB than LFW, while AHF of starch was 7.5 times greater for HFAB than LFW. The calculated NE value was greatest for LFW, intermediate for HFBB and HFAB, and lowest for LFB. Among diets, HFAB increased the concentration of butyrate and propionate in cecal and colonic digesta, with butyrate concentration being up to 50% more than feeding LFW diet. Carcass protein and lipid content did not differ among pigs fed the 4 diets.

Implications: Feeding barley instead of wheat grain to weaned pigs reduced diet nutrient digestibility but it did not affect carcass protein and lipid content. Feeding high amylose barley increased gut microbial SCFA production, potentially supporting better gut immune barrier function and gut development, thereby benefiting the gut and overall health of weaned pigs.

Response of growing pigs to the inclusion of hybrid rye in low or high energy diets

Miranda N. Smit[1], Josi C. Panisson[1], and A. Denise Beaulieu[2]

[1]Prairie Swine Centre Inc., Saskatoon, SK, S7H 5N9; [2]Department of Animal and Poultry Sciences, University of Saskatchewan, Saskatoon, SK, S7N 5A8.
Email: miranda.smit@usask.ca or denise.beaulieu@usask.ca

This experiment was designed to determine if increased dietary energy content could compensate for the reduced feed intake previously shown in pigs fed diets with hybrid rye replacing wheat. We hypothesized that pigs fed 40% hybrid rye would perform better on the high than the low energy diets. A total of 160 pigs (70.1 kg BW) housed in 32 pens, 5 pigs/pen, were fed diets with 0 or 40% hybrid rye (var. 'KWS Bono') replacing wheat, either with low (2,350) or high (2,450) kcal NE per kg of diet over two growth phases (Phase 1; 70-85 kg BW; Phase 2; 85-130 kg BW). The BW and feed disappearance were measured. Fecal samples obtained in Phase 2 (~100 kg BW) were used to calculate apparent total tract digestibility (ATTD) of gross energy (GE). The hybrid rye fed in this experiment had relatively low levels of mycotoxins, including ergot. The ATTD of GE was unaffected by rye inclusion (average 80.5%) and was reduced ($P = 0.001$) in the low vs. the high energy diets (79.4 vs 81.6%, respectively). In the first 17 days, pigs fed the low energy hybrid rye diets gained less weight ($P < 0.01$) than those fed the other three diets, but final BW and overall (d 0-50) feed intake were not affected by rye inclusion or NE level. The NE intake was greater ($P < 0.05$) and feed efficiency (G:F) was reduced ($P < 0.05$) in pigs fed rye diets compared to those fed diets without rye, whereas there was no effect of NE level on NE intake or G:F. The low NE diets were cheaper ($P < 0.01$) than the high NE diets, which translated to $5.22 lower feed cost per pig and $0.06 lower feed cost per kg BW gain for pigs fed low vs. high NE diets. There was no effect of rye inclusion on feed cost per pig, but feed cost per kg BW gain tended ($P = 0.059$) to be 0.04 CAD higher for diets with rye compared to those without due to the lower overall feed efficiency for pigs fed the hybrid rye diets.

Implications: The inclusion of 40% hybrid rye in diets for growing finishing pigs resulted in only minor changes in growth performance. Increasing the NE level of the first phase diet in the grower-finisher barn may be useful to avoid a reduction in growth performance when feeding hybrid rye. However, reducing the dietary NE level can be a good strategy to save money on feed costs in the grower-finisher barn.

Fermentability of barley endosperm and processed barley hulls affect nutrient digestibility in weaned pigs

L. Zhu[1], L.F. Wang[1], R.N. Newkirk[2], A.D. Beattie[2], E. Beltranena[1], and R.T. Zijlstra[1,3]

[1]University of Alberta, Edmonton, AB, Canada T6G 2P5; [2]University of Saskatchewan, Saskatoon, SK, Canada S7N 5A8; [3]Email: zijlstra@ualberta.ca

Barley contains non-digestible fiber that reduces apparent total tract digestibility (ATTD) of energy (GE) and protein CP). However, cereal grain fiber can benefit the animal gut through fermentation and may reduce post-weaning diarrhea and enhance growth. The endosperm of different barley cultivars varies in digestibility. Moreover, steam explosion may open fiber structures and increase the digestibility of fiber in barley hulls. To explore, six diets containing either 539 or 614 g/kg of low- or high-fermentable barley endosperm, plus 80 or 91 g/kg of three hulls: 1) oat, 2) untreated barley, 3) steam-exploded barley; in phase 1 or phase 2, respectively. Diets were cold-pelleted for phase 1 and steam-pelleted for phase 2. Diets were fed to 216 weaned pigs starting 1-week after weaning for 4 weeks in 2 phases (14 days each). For phase 1, low fermentable endosperm combined with oat or untreated barley hull, but not steam-exploded barley hull, had greater ATTD of GE and CP than high fermentable endosperm. For phase 2, the low fermentable endosperm had greater diet ATTD of GE and CP than high fermentable endosperm. Steam-exploded barley hull had greater diet ATTD of GE, but not ATTD of CP, than untreated barley hull. The greater nutrient digestibility of the low fermentable barley endosperm may contribute to the greater growth of pigs that we reported previously.

Implications: Weaned pigs have better feed intake and growth when fed diets containing less fermentable carbohydrates than with diets containing more fermentable carbohydrates. Furthermore, steam-explosion of barley hull may increase energy digestibility.

Effects of acid-binding capacity in low-zinc diets on growth performance of weaned pigs

J. L. Landero[1], L.F. Wang[2], L. Zhu[2], J.R. Bergstrom[3], M.G. Young[1], and R.T. Zijlstra[2,4]

[1]Gowans Feed Consulting, Wainwright, AB, Canada; [2]University of Alberta, Edmonton, AB, Canada, [3]DSM Nutritional Products, Parsippany, NJ, USA; [4]Email: ruurd.zijlstra@ualberta.ca.

Post-weaning diarrhea of piglets has historically been controlled with dietary solutions such as antibiotics as growth promotors and high dietary inclusion of ZnO. However, such use of antibiotics is not permitted any longer and such use of ZnO has already been banned in the EU. In the present trial, the objective was to study the effects on growth performance of acid binding capacity (ABC) as a dietary solution in low zinc diets and compare to a control diet with high zinc. In total, 1445 pigs (Camborough F1 female x 800 boar line; PIC Canada) were weaned at 20 ± 1 d of age, and were randomly placed into one of 50 mixed-gender pens (50% barrows and 50% gilts) in two rooms filled four days apart, for 28 or 29 pigs per pen. In total, five dietary regimes were tested during Phase 1 (day 0-7) and 2 (day 7-21): one high in Zn (2,000-3,000 ppm ZnO; 412 mEq/kg ABC-4) and 4 low in Zn (150 ppm) with incremental reductions in ABC-4 from 326 to 150 mEq/kg. Reductions in ABC-4 were accomplished using a combination of changes: added organic acids (formic, fumaric, benzoic), Ca formate to replace limestone, and a low ABC-4 soy protein concentrate to replace regular soy protein concentrate. Pigs were fed a common diet during Phase 3 (days 21-42). For each feeding phase all diets were formulated to equal net energy and standardized ileal digestible amino acids and contained phytase. At day 21, pigs fed high Zn were 1.6 kg heavier (11.3 vs 9.7 kg) than pigs fed low Zn with highest ABC-4. Within low Zn diets, reducing ABC-4 linearly increased body weight (9.7 vs. 10.1 kg). These differences continued until day 42: pigs fed high Zn were 1.8 kg heavier (24.1 vs 22.3 kg) than pigs fed low Zn with highest ABC-4). Within low Zn diets, reducing ABC-4 for the first 3 weeks after weaning linearly increased body weight (22.9 vs. 22.3 kg). This increased growth was caused by a combination of increased feed intake and increased gain-to-feed and coincided with reduced incidence of scours.

Implications: Weaned pigs fed high Zn diets grew more than pigs fed low Zn diets. Dietary ABC-4 is a dietary solution to consider for ameliorating the reduced growth by pigs fed low Zn diets: 33% of the reduced growth could be compensated for by feeding diets low in dietary ABC-4 during the first 3 weeks after weaning.

Disease resilience indicators based on immune stimulation of blood from young healthy pigs

V. Bhatia[1*], X. Bai[2,3], F. Fortin[4], J.C.S. Harding[5], M.K. Dyck[3], C.J. Field[3], C. Rogel-Gaillard[6], F. Blanc[6], G.S. Plastow[3], PigGen Canada[7], and J.C.M. Dekkers[1]

[1]Dept of Animal Science, Iowa State University, Ames, IA; [2]Hendrix Genetics, Boxmeer, Netherlands; Livestock Gentec, Dept. of Agri., Food and Nutritional Science, University of Alberta, Edmonton, AB; [4]Centre de Développement du Porc du Québec, Inc., Quebec City, QC; [5] Dept. of Large Animal Clinical Sciences, University of Saskatchewan, Saskatoon, SK; [6]Université Paris-Saclay, INRAE, AgroParisTech, GABI; [7]PigGen Canada, Guelph, ON.
Email: vishesh@iastate.edu

In pig production, infectious diseases pose significant challenges and dramatically impact production efficiency and animal welfare. Breeding for disease resilience, defined as an animal's ability to maintain production performance under disease, is an option but collecting disease resilience phenotypes in disease-free nucleus breeding herds is challenging. This study investigated immune response phenotypes measured on young, healthy animals as potential indicators to select for disease resilience in high-health nucleus breeding herds. Samples and data used were from a wean-to-finish natural disease challenge model, in which young healthy Landrace × Yorkshire barrows from one of seven breeding companies were placed in a quarantine nursery before exposure to multiple diseases. Blood immune stimulation assays for innate and adaptive immune responses were evaluated. The assays measured cytokine concentrations and mean fluorescence intensity in blood samples collected at around 26 days of age from 1524 animals in the quarantine nursery, following one of three treatments of the samples (Non-Stimulated, LPS, and PI). Heritabilities and genetic correlations with disease resilience phenotypes collected on these pigs and on an additional 2571 pigs were estimated. Results showed blood stimulation assay measures to be moderately heritable (average 0.08; range 0 to 0.27), while estimates of genetic correlations with disease resilience traits ranged from -0.75 to 0.83, but with substantial standard errors. This project was supported by Genome Canada, Genome Alberta, PigGen Canada, and USDA-NIFA grant #2017-67007-26144.

Implications: This study provides insights into blood stimulation indicator traits (IL-2,IL-4, and IL-17) that have potential to be utilized to select for disease resilience, enabling pigs to maintain production performance under disease.

Whole bacterial cell ELISA for relative measurement of *Streptococcus equi* subsp. *zooepidemicus* antibody levels in serum

Hannah Burlet[1], Ana Norte[1], Roman Nosach[1], Arthur Finatto[1], Daniela Gutierrez Montiel, Suzana Stubert, Alison Jeffery[1] and Matheus Costa[1,2]

[1]University of Saskatchewan, Saskatoon SK; [2]Utrecht University, Utrecht, NL
Email: hannah.burlet@usask.

Streptococcus equi subsp. *zooepidemicus* (*Strep zoo*) is a swine pathogen of increasing concern in North America and worldwide. This disease is characterized by septicaemia, and sudden death in pigs of any age, and abortion in sows. Enzyme-linked immunosorbent assay (ELISA) is a sensitive test that allows for relative measurement of antibody levels in a given sample. The most common ELISA tests measure antibody levels against a single antigen. However, isolating a specific antigen for coating is both costly and time consuming. Using whole bacterial cells as antigen avoids this step while maintaining specificity. The goal of this study was to develop a whole cell ELISA for confirmation and relative measurement of *Strep zoo* seropositivity in swine. Ninety-six well plates were coated with killed whole *Strep zoo* ST-194 grown to 10^8 CFU. After coating, drying, and fixing, plates were washed with PBS-Tween 0.05% (PBST). Plates were blocked with skim milk. Unknown serum was diluted in PBST. Positive and negative controls were included, as well as a no-sample control. Plates were washed after 1 hour incubation. Goat anti-swine conjugated IgG was added and incubated for 1-hour and washed. To develop a colorimetric reaction TMB was added, and the reaction was stopped when positive controls reached an optical density of 0.5 at 450 nm. Clinical samples (n=16) from pigs prior to and 21 days following exposure to *Strep zoo* were evaluated. Antibody titer was calculated based off absorbance and compared to positive and negative controls. We anticipate that antibody titers 21 days post-exposure will be significantly higher than pre-exposure samples.

Implications: A serological test capable of discerning subclinical pigs from non-infected pigs may support biosecurity measures to prevent the inadvertent introduction of *Strep zoo* carrier pigs to naïve herds.

Assessing the Prevalence of *C. Difficile* and *Clostridium perfringens* type A on Canadian Farms with Enterocheck

DeGroot, M.*; Boix, O.; Galé, I. and Ibanez, G.

HIPRA, Amer (Girona), Spain.
Email: mike.degroot@hipra.com

Neonatal diarrhea (ND) is a common occurrence on Canadian pig farms. The cause of neonatal diarrhea in piglets is often multifactorial; many infectious pathogens can contribute to diarrhea. This includes *E. coli*, *Clostridium perfringens* (type A and C), *Clostridioides difficile* (*C. difficile*), Rotavirus, and Porcine Epidemic Virus (PED). However, the prevalence of *C. difficile* and *Clostidium perfringens* type A (CpA) on Canadian pig farms has not been recently investigated. *C. difficile* causes diarrhea in piglets less than 7 days of age, with high morbidity and variable mortality. CpA is a normal inhabitant of the pig's intestinal microbiota and has also been implicated as a cause of diarrhea in neonatal piglets. Understanding the prevalence of *C. difficile* and CpA would allow for better control of ND on Canadian farms. Piglets on 54 Canadian farms with ND were sampled with rectal swabs and samples were placed on FTA cards. Testing was done by using Enterocheck kits, an FTA-card diagnostic kit based on polymerize chain reaction (PCR) testing. Cycle threshold (Ct) values were determined, and the presence of A and B toxins of *C. difficile* were considered relevant when Ct value <30 and α-toxin of CpA was considered relevant with Ct-value <26. 84.94% of the farms tested positive for *C. difficile* A toxin, with high positivity (Ct<30) on 41.57% of farms. In addition, 31.93% of the farms tested positive for *C. difficile* B toxin, with a high positivity on 2.41% of farms. For *C. perfringens*, 100% of the farms tested positive for α toxin, with high positivity (Ct<26) on 69.28% of farms.

Implications: This study demonstrates that *C. difficile* and CpA are both highly prevalent on Canadian pig farms. Also, with high positivity *C. difficile* and CpA need to be considered as potential primary pathogens in cases of ND. Identifying *C. difficile* and CpA and implementing control strategies, such as vaccination, can help reduce ND on Canadian farms.

Refining an animal model for porcine ear-tip necrosis

Alison Jeffery[1], Hannah Burlet[1], Roman Nosach[1], Matheus Costa[1,2]

[1]Large Animal Clinical Sciences, Western College of Veterinary Medicine, University of Saskatchewan, Saskatoon, SK.
[2]Utrecht University, Faculty of Veterinary Medicine, Utrecht, NL.
Email: alison.jeffery@usask.ca

Ear-tip necrosis (ETN) is an infectious disease that results in progressive loss of the ear in nursery and grower pigs. Lesions start at the tip of the ear, as reddening and edema, and may progress to complete necrosis. This condition has been reported in all pork-producing countries. ETN is a welfare concern as affected pigs may become disfigured, losing parts of one or both ears. It is also a seemingly painful condition. Wounds in severe cases may provide a port-of- entry for other pathogens, leading to secondary infections that may negatively affect performance. The etiology and pathogenesis of ETN is unclear. There is no effective treatment available. Our group has replicated ETN by exposing pigs to a single infectious agent. The objective of this study was to further advance and refine our previous challenge model. This study aimed to evaluate microneedles as an inoculation device to replicate ETN. Eight 5-weeks old pigs were housed at a biosafety level 2 facility. Two pigs served as sentinels, controlling for spontaneous disease development and were not challenged. The right ear of the remaining 6 pigs were inoculated using microneedles, left ears were not inoculated. Animals were clinically monitored twice daily for 7 days for the development of lesions. Results will be presented at the meeting.

Implications: Development of an animal challenge model will further aid in profiling disease pathogenesis, progressing farm management practices and the development of a non-antibiotic treatment.

Transepithelial ion transport in the stomach of pigs exposed to gastric ulcer conditions

Agbee L. Kpogo[1], Josiane C. Panisson[2], Denise Beaulieu[1], Rex Newkirk[1], and Matthew Loewen[3]

[1]Department of Animal and Poultry Science, University of Saskatchewan, Saskatoon, SK, S7N 5A8; [2]Prairie Swine Centre, Inc., SK, S7H 5N9 ; [3]Department of Veterinary Biomedical Sciences, Western College of Veterinary Medicine, University of Saskatchewan, Saskatoon, SK, S7N 5B4
Email: agbee.kpogo@usask.ca

Gastric ulcers in pigs increase significantly when stressful events occur. However, the exact causes of these gastric ulcerations remains unclear. Gastric acid production is known to be influenced by anionic secretion. This study aimed to evaluate the agonist-induced electrogenic secretory responses in the different sections of the stomach of pigs exposed to gastric ulcers. The *pars oesophagea*, cardia, fundus and pylorus were assessed in Ussing chambers. Pigs were fed diets with 40% air-classified pea starch (particle size 10-20 μm) and exposed to an out-of-feed event, both recognized as potential factors contributing to gastric ulcer formation. Sixteen pigs (initial BW of 90.6±2.2 kg), were fed for 14d and divided into 4 treatment groups (0h, 24h, 48h, 72h). On d13, feed was removed from 7am and reintroduced on d14 at 7am. The 0h treatment had feed throughout this period. After the "feed out" event, 4 pigs each from the 24 h treatment groups were selected and euthanized after 24h, 48h and 72h following feed reintroduction. Changes in short-circuit current (I_{sc}) in the 48h pigs significantly decreased following the activation of secretion via the adrenergic agonist (isoproterenol) in the cardia, and the cholinergic agonist (carbachol) in the *pars* and pylorus. There were no significant secretory responses in the other segments. Since bumetanide failed to inhibit the basolateral cotransporter 1 (NKCC1), we attribute the changes in the stimulatory I_{sc} to bicarbonate (HCO_3^-) secretion. We report for the first time that the pars oesophagea becomes an actively secreting tissue during gastric ulcer conditions. Also, the 48h group showed the highest secretory response and the highest ulcer scores. We propose that HCO_3^- may play a pivotal role in gastric ulcer conditions and further studies such as qPCR and the inclusion of Acetazolamide as an agonist in the Ussing chamber studies are imperative to proof this concept.

Implications: Understanding the physiological changes and contributions towards gastric ulcers in pigs will help scientists fully understand the etiology of gastric ulcers in pigs and be useful in the development of drugs.

A blend of spice oleoresins can substitute the growth promoting effect of salinomycin on grower-finisher hogs: A commercial evaluation

Bertrand Medina[1]*, Dan Bussières[2] and Ivan D. Girard[1]

[1]Probiotech International Inc., St-Hyacinthe, Québec J2S 8L2, Canada; [2]Groupe CÉRÈS Inc., St-Nicolas, Québec, G7A 3S8, Canada
*Email : bertrandmedina@probiotech.com

The swine industry is worried about the plight to rethink the use of antibiotic agents for mitigating any risk of antimicrobial resistance and thus responds to consumers' health concerns. The aim of this experiment is to evaluate a proprietary blend of spice oleoresins (CAPS C2PG) as an alternative to salinomycin-growth promotor in growing-finishing pigs. 704 pigs (*Improvest*® injected males) were housed in 32 pens distributed in 4 rooms (8 pens/room) of a compound building fully equipped with an automatic and computerized *Gestal*™ Feed system. During the grower and finisher phases, the pigs were allotted to three different treatments (16 repetitions/treatment): salinomycin (25 ppm) (Sal.); CAPS C2PG at 35 g/MT (CAPS.). Growth performances (ADG) are reported based on pen weight at entry and then every 3 weeks and feed intakes are daily monitored per pen. All dead or culled pigs are weighed upon removal. Data are analyzed by an ANOVA model including treatment and pen as fixed and random effects, respectively. Pairwise comparisons are completed with a Chi-square test. Under these commercial conditions, the overall performances (ADF, ADFI and FCR) do not show any significant differences between treatments. The proprietary blend of spice oleoresins represents an alternative solution to substitute the growth promoting effect of 25 ppm of salinomycin during the fattening phases of male pigs. From an economic balance, the cost of 35 g/MT of CAPS C2PG feed additive is around $1.50/MT of feed, which is below to the cost of 25 ppm of salinomycin ($1.75-2.00/Ton).

Implications: Some economic, societal, sustainable and health advantages exist to substitute an ionophore growth promoter with an alternative and natural solution based on spice oleoresins.

Clinical benefits of an in-water essential oil blend on antibiotic-free pigs experiencing post-weaning diarrhea: A field study

Gillian Greaves[1], Clint Lichty[1], Bertrand Medina[2*] and Ivan D. Girard[2]

[1] SouthWest Ontario Veterinary Service, Stratford, Ontario N4Z 1H3, Canada; [2] Probiotech International Inc., St-Hyacinthe, Québec J2S 8L2, Canada
* Email: bertrandmedina@probiotech.com

Post-weaning diarrhea can be costly for producers, especially in antibiotic free production flows (RWA), where the use of antimicrobials leads to a loss of valuable premiums on individual pigs. This field study examined the impact of an in-water essential oil blend on RWA pigs with post-weaning diarrhea using pen-level diarrhea scores, and percentages of diarrhea-related sick, fallback and removed pigs. A total of 3,598 pigs (~4 weeks of age, mixed sex) in 73 pens from 2 RWA flows and 4 pig turns (2 turns per nursery barn) were enrolled. Sow sources were known to experiencing mild *E. coli* and/or rotavirus associated post-weaning diarrhea. In each barn, half of the pens were assigned to control (regular water) or treatment [Entero-V® Swine, at 250 mL /1000L drinking water. and supplying Star Anise, Turmeric and Cinnamaldehyde oil blend] for ~2 weeks post weaning. Pen-level diarrhea was assessed every other day (scale from 0 to 4). Additionally, diarrhea-related sick pigs were assessed individually at the same time pen-level diarrhea was scored. In each pen, a pig was first identified as sick (sunken eyes, poor gut fill or lethargic/depressed) and then confirmed if it had diarrhea. Diarrhea-related fallbacks were assessed at the end of treatment (~ 2 weeks post-weaning). A pig was considered fallback if it did not recover from its enteric problem (previously marked as a diarrhea-related sick pig) by the end of the treatment. A diarrhea-related removal was a previously marked diarrhea-related sick pig that was removed from its pen. Mild post-weaning diarrhea related to *E. coli* and rotaviruses occurred in both flows. Overall, the VHP reduced pen-level diarrhea scores over time (1.2 *vs* 1.5, p <0.01), and percentages of diarrhea-related sick (10.9 *vs* 14.3%, p=0.36), fallback (3.5 *vs* 4.6%, p=0.10) and removed pigs (0.17 *vs* 0.30%, p=0.12) compared to the control group

Implications: These first results suggested that an in-water essential oil blend could be an aid in reducing clinical impact of mild post-weaning diarrhea caused by *E. coli* and rotavirus in RWA pigs.

Comparing motility, biofilm formation, and Amoxicillin resistance of *E. coli* isolated from wild and conventional pigs

Joshua Theriault[1], Rajibur Rahman[1], Tingting Ju[1], Janelle M. Fouhse[1], Julia Neyedly[1], Ryan K. Brook[2], and Benjamin P. Willing[1]

[1]AFNS, University of Alberta, AB, Canada. [2]University of Saskatchewan, SK, Canada.

Antibiotic-resistant bacteria, particularly *Escherichia coli*, is a significant threat to global public health. Recent research shows that the flagellar motility of *E. coli* is critical for biofilm formation and antibiotic resistance. However, there is a lack of information regarding motility, biofilm formation, and antibiotic resistance of *E. coli* isolated from pigs. Therefore, we characterized and compared *E. coli* isolates collected from conventional sows (n=96), piglets (n=12) and wild boar (n=34) for motility, biofilm formation, and antibiotic susceptibility. All the isolates were investigated for amoxicillin susceptibility through the dilution method. Bacterial motility was measured using tryptone yeast extract soft agar. The biofilm-forming potential based on curli protein was initially screened using Congo red indicator media and subsequently quantified via a modified air-liquid interface technique. Results indicated that all wild boar isolates were motile, and only 8.8% were resistant to amoxicillin. In contrast, *E. coli* strains isolated from conventional pigs showed different phenotypes between sows and piglets. Most (83.3%) of the isolates from conventional piglets were resistant to amoxicillin, and 16.7% were motile. However, in conventional sows, nearly half of the isolates (48%) were amoxicillin-resistant, and 35% were motile. In addition, wild boar isolates (67.6%) showed a higher presence of curli protein, followed by sow (52.1%) and piglets (25%), indicating a greater potential of wild boar isolates to form biofilm than conventional pig isolates. Interestingly, compared with the normal growth condition, amoxicillin-resistant *E. coli* strains showed increased motility (P = 0.01) and biofilm formation (P = 0.04) with amoxicillin supplementation. We found that wild boar *E. coli* strains were more motile, better biofilm formers, and less amoxicillin-resistant than those isolated from conventional pigs.

Implication: Our research demonstrated unique characteristics of commensal bacteria in different farming systems and environments, which provides insight into bacterial adaptation to selective pressures and how we can improve production environmental management for antimicrobial resistance.

Effects of rye bran as a feed component in the diets of lactating sows on sow and piglet performance

Christian Homann[1], Volker Wilke[1], Isabell Eckey[1], Andreas von Felde[2] and Christian Visscher[1]

[1]Institute for Animal Nutrition, University of Veterinary Medicine, Foundation, 30559 Hannover, Germany; christian.homann@tiho-hannover.de (C.H.); volker.wilke@tiho-hannover.de (V.W); secretariat-tierernaehrung@tiho-hannover.de (I.E); christian.visscher@tiho-hannover.de (C.V.)
[2]KWS Lochow GmbH, 29303 Bergen, Germany; andreas.vonfelde@kws.com

From a cost and sustainability perspective, the use of food processing by-products such as rye bran in sow diets is becoming increasingly attractive. In general, rye bran has a high content of dietary fiber, vitamins, minerals, and other bioactive substances. Of particular interest is the high content of non-starch polysaccharides (NSP), which lead to increased fermentation to short-chain fatty acids, especially butyrate. Butyrate is said to have many positive effects on the intestinal health of pigs. The effects on performance and acceptance were examined in this study. A commercial compound feed containing wheat, barley, soybean meal and wheat bran (9.0%) was used as control diet (C). On this basis, an isonitrogenous and isoenergetic experimental diet (R) with 15% rye bran was calculated. In four runs, a total of 80 sows were weighed and backfat thickness was measured when entering and leaving the farrowing pen. Moreover, suckling piglets from 80 litters were weighed at birth and on days 1, 7, 14 and 20 of life to determine daily gain. Estimation of colostrum intake of the piglets and milk production of the sows were determined by established methods (Theil 2014, GfE 2006). The average daily gain (g) of the suckling piglets in the experimental group showed no significant difference compared to the control group (C 196.50±33.22, R 198.91±27.00). Moreover, no significant differences regarding body weight loss (kg) (C 30.17±12.27, R 28.31±12.78) and the loss of backfat thickness (mm) (C 2.25±1.29, R 2.18±2.11) of the sows occurred during lactation. When comparing the estimated colostrum intake of the suckling piglets per sow (g) (C 4629.58±785.15, R 4770.06±825.48) and the average daily milk production of the sows during lactation (kg) (C 11.36±1.64, R 11.83±1.41), no significant differences were found either.

Implications: The results of the study showed that 15% rye bran can be used as a feed component in the diets of lactating sows without negative effects on piglet and sow performance. Further studies are needed to demonstrate possible positive effects in piglet production.

Circulating biomarkers differ in sows with variable risk for pelvic organ prolapse

Zoë E. Kiefer, Jamie M. Studer and Jason W. Ross

Department of Animal Science, Iowa State University
Email: jwross@iastate.edu

A substantial portion of sow mortality is caused by pelvic organ prolapse (POP) during late gestation and early lactation. Despite being widely acknowledged as a sow welfare and production issue, the underlying causative mechanisms are not well understood. Previous studies have demonstrated biomarkers within circulation differ between sows with differing risk for POP. Additionally, analytes and metabolites in serum have also been observed to be different in sows at differing risks for POP as well as sows that subsequently prolapsed compared to those that did not. The objective of the current study was to further evaluate circulating biomarkers in sows with presumed high risk for POP compared to sows with presumed low risk for POP during late gestation as well as between sows that did or did not subsequently experience POP. We hypothesized that circulating biomarkers would differ between sows at varying risk for POP. During late gestation (days 110-115), sows (n = 108) from two different farms were evaluated for POP risk using a 3-point perineal scoring system. Briefly, a perineal score (PS) of 1 (PS1) presumes little to no risk of POP, a PS of 2 (PS2) presumes moderate risk of POP, and a PS of 3 (PS3) presumes high risk of POP. Blood samples were collected from PS3 sows (n = 58) and parity matched PS1 sows (n = 50). Glucose concentration was determined using the Abaxis VetScan VS2 benchtop analyzer and the Kidney Profile Plus prepackaged rotor. Additionally, insulin and β-Hydroxybutyrate (BHB) were measured using commercially available kits according to manufacturers' instructions. Statistical analysis was conducted to evaluate effects of PS, Farm, the interaction of PS and Farm, and prolapse outcome. Glucose concentration was not affected by PS or prolapse outcome, although a difference in glucose between farms was observed ($P = 0.03$). Insulin was 32.7% lower ($P = 0.04$) in PS3 sows compared to PS1. A tendency ($P = 0.10$) for a higher glucose to insulin ratio was observed in sows that subsequently experienced POP compared to sows that did not. Additionally, BHB was 15.8% lower ($P = 0.04$) in PS3 sows compared to PS1, and 17.2% lower ($P = 0.04$) in sows that subsequently experienced POP compared to those that did not. Collectively, these data indicate sows at high risk for POP have differences in serum biomarkers compared to sows at low risk, and differences exist between sows that did or did not subsequently experience POP. Further investigation is needed to better understand how these differences relate to POP.

Implications: Prolapses account for 25% of sow mortality and by reducing POP incidence we can improve productivity of the swine industry as a whole.

Identification of small molecules in sows affected by postpartum dysgalactia syndrome

Jamie M. Studer, Zoe E. Kiefer, Lance H. Baumgard, Stephan Schmitz-Esser, Laura L. Greiner, Aileen F. Keating and Jason W. Ross

Department of Animal Science, Iowa State University, Ames, IA
Email: jwross@iastate.edu

Management of hyperprolific sows is a metabolic and physiological challenge for producers, as an increased litter size has elevated metabolic demands of gestation and lactation. The imbalance between nutrient consumption and energy expenditure during the transition period creates a catabolic state, as energy is used for fetal growth, mammary development, and milk synthesis. Negative energy balance induces oxidative stress, insulin resistance, and may cause intestinal barrier dysfunction, leading to chronic low-grade inflammation and rendering sows vulnerable to disorders such as postpartum dysgalactia syndrome (PDS). Described as a metabolic and hormonal disturbance, PDS is characterized by insufficient production of colostrum and milk which compromises piglet growth and increases pre-wean mortality. Therefore, objectives were to utilize liquid-chromatography mass-spectrometry (LC-MS) to identify alterations in serum small molecules and pathways associated with PDS in sows. To accomplish this, a retro-classification study was conducted on a commercial sow farm with poor lactation performance. Thirty-six sows with clinical signs of PDS were matched by parity and day of lactation with thirty-six healthy control (CON) sows. Serum samples collected at farrowing (timepoint 1) and at detection of PDS (timepoint 2) were subjected to LC-MS. Evaluation of the metabolome revealed differences in 72 and 145 molecular features between CON and PDS sows at timepoint 1 and timepoint 2, respectively ($P < 0.10$). Metabolite enrichment analysis revealed glycerophospholipids were enriched at both timepoints, with 8 and 20 metabolites identified as glycerophospholipids at timepoint 1 and timepoint 2, respectively. As major components of biological membranes, glycerophospholipids are responsible for regulating stability and permeability of the lipid bilayer. Phosphatidylcholine (PC) and phosphatidylethanolamine (PE) are the most abundant phospholipids in cell membranes, and changes in PC and PE levels have been implicated in metabolic disorders. Alterations in phospholipid levels may be associated with parameters related to maternal energy balance such as lipid profiles and insulin resistance.

Implications: Understanding the serum metabolome could provide insight into the pathophysiology underlying the development of the multifactorial disease complex known as PDS. Furthermore, small molecules altered in PDS sows at timepoint 1 could serve as biomarkers to aid in early detection of PDS.

The use of uterine horn resection surgery to optimize non-surgical embryo collection in gilts

M. E. Ewasiuk[1], R. R. E. Uwiera[1], A. I. D. Woon[1], L. J. Zak[2], and M. K. Dyck[1]

[1]Department of Agricultural, Food, and Nutritional Science, University of Alberta, Canada, T6G 2P5; [2]Topigs Norsvin, Den Bosch, Netherlands.
Email: mkdyck@ualberta.ca

Embryo transfer has proven to be a financially beneficial practice across a multitude of livestock sectors including cattle, horses, sheep, and goats, hence its demand in the swine industry. However, the convoluted nature of the pigs' uterine horns, ranging between 1.5-2m in length, and the development of an appropriate catheter pose as great challenges to standardizing such a procedure. A surgical resection approach provides a transitional method to navigate these barriers by improving accessibility to the utero-tubal junction for ease of embryo retrieval. The optimization and implementation of non-surgical embryo collection (nsEC) techniques would aid the investigation of embryo survival, quality, and embryonic interactions within the uterus. In conjunction with embryo transfer, these methods offer a superior means of disseminating swine genetics globally while protecting herd health and eliminating animal welfare concerns associated with live animal transport. In this study, 20 purebred gilts (parity zero) underwent a multi-step uterine horn resection surgery to simplify and shorten the female genital tract. Following recovery from surgery and cervical artificial insemination, the reproductive tract was flushed in vivo between days 4 to 7 following the onset of estrus. The rate of embryo recovery was 21.7% following 60 nsECs. Despite an average media recovery of 60.7% (range: 11-100%; median: 65.7%), the number of embryos collected ranged between 1-5. Further research must focus on the development of a refined nsEC catheter and tip that can successfully and repeatedly navigate the complex reproductive tract of a parity one sow with high recovery of flushing media and therefore, embryos.

Implications: Collecting pre-implanted porcine embryos through nsECs would provide an avenue for nuanced embryonic analyses to in turn improve the efficiency and profitability of breeding stock herds. These benefits include facilitating the transport of genetically superior embryos in a sterile and timely manner without posing any risk to farm biosecurity or animal welfare.

Feed restriction in finisher pigs alters their activity levels and leads to a decrease in the occurrence of tail biting

Kayla Patey[1], Jennifer Patterson[1], Michael Dyck[1], Clover Bench[1], Graham Plastow[1], Elda Dervishi[1*]

[1]Department of Agriculture, Food and Nutritional Science, Faculty of Agricultural, Life and Environmental Sciences, University of Alberta, Edmonton, AB, Canada, T6G 2P5. *dervishi@ualberta.ca

The study investigated the impact of early growth feed restriction on pigs' activity and tail-biting behavior. Conducted at the Swine Research and Technology Centre, University of Alberta, it involved 132 crossbred grower-finisher pigs distributed over 3 cycles. At 8 weeks of age 48 pigs/cycle were allocated in 8 pens, 6 pigs/ pen, balanced for sex, 3 males 3 females, and weight. Half of the pens received 80% of the recommended feed allowance (RD) for 3 weeks from week 9 of age and half of the pigs received the full diet (FD). Pigs were then fed the recommended diet until the end of the experiment. Pigs were recorded nine times, once per week at week 8, 9, 10, 11, 12, 13, 15, 18 and 20 of life using EthoVision XT (Noldus), 4 hours in the morning and 4 hours in the afternoon. For each area of the pen (feeders, drinkers, central area, left and right corner), several activity variables were extracted including: total distance moved (cm), frequency (times), cumulative duration (s), latency to first visit (s), mean velocity (cm/s) and mean mobility (%). The recording was used for the observation of the tail biting behavior. The number of scans where pigs were observed to perform or received tail biting was used to classify the pigs in performer, receiver, neutral and performer-receiver. A generalized linear mixed model (R version 4.3.0) was fitted to test the effect of feed restriction on the number of scans where pigs were observed to perform or receive tail biting. A linear mixed model with repeated measurement was fitted to test the effect of pig classification on the activity variables. Feed restriction at week 9-11 of age was associated with less tail biting. During the feed restriction period, pigs increased their number of visits to the feeders and close to feeder areas ($P<0.05$). At 16, 18 and 20 weeks of age, pigs that performed tail biting, visited the water area more frequently and spent more time at the drinkers ($P<0.05$). This study shows that 20% feed restriction at weeks 9-11 of age affects pig behavior and reduces the number of tail biting performed and received.

Implications: This study provides new information to potentially help develop feeding and management strategies for the industry to better manage tail biting episodes on their farms.

Measuring stress resilience in an out-of-feed event using cortisol and oxytocin in growers given early life management modifications

S. Khalife*[1], D. Pollock[1], M. Lopez-Arjona,[2] J.J. Ceron[3], J.A. Brown[4], and Y.M. Seddon[1]

[1]Western College of Veterinary Medicine, University of Saskatchewan, Saskatoon, Canada, S7N 5B4. [2]Universitat Autònoma de Barcelona, 08193 Cerdanyola del Vallès, Barcelona, Spain. [3] Faculty of Veterinary Medicine, Murcia University, 30100 Murcia, Spain. [4]Prairie Swine Centre, Saskatoon, Canada, S7H 5N9.
*Email: siba.khalife@usask.ca

Pigs undergo rapid development pre-weaning (0-4 weeks), and during nursery/grower (4-12 weeks) where conditions experienced impact the development of stress responses with lasting impacts. This study investigated if targeted early life management modifications, designed to modulate stress resilience, can impact the response of grower pigs to a production relevant stressor, an out-of-feed event. Litters were assigned at birth to enhanced (E;n=20 litters, mean: 8 pigs/litter) or standard (S; n=20 litters, mean: 10 pigs/litter) treatments. E pigs received three, simultaneous management modifications: extra space (E:6m^2 vs S:4.32m^2) to promote social development, chewable materials (rope, burlap) to shape oral behaviour, and human contact (5-min, 3 days/week) to reduce fear of humans. At weaning (26±2 days, mean±SD), E and S pigs were kept on the same (Enhanced-Enhanced, Standard-Standard), or switched to the opposite (Enhanced-Standard, Standard-Enhanced) treatments. At 12 weeks, pigs returned to standard conditions and were followed to slaughter (130kg). At 125 days, feed was cut off for 20 hours (day 1 at 4:00pm, till day 2 at 12:00 pm). Saliva was taken 1 hour before (T1), 16 hrs after feed removal (T2), and 1 hour after (T3) feed re-introduction with behaviour & skin lesions taken. Saliva was analyzed for cortisol (physiological stress) and oxytocin (emotional well-being). Treatments did not differ in salivary hormones and a high variation between pigs was observed at all time points for cortisol (T1:205±129, T2:171±95, T3:194±111 ng/mL, median±SD) and oxytocin (T1:553±1817, T2:455±1155, T3:513±1500 pg/mL).

Implications: There was high variability in hormone concentrations across treatments, with no difference between treatments. Further analysis of behaviour, injury, and two more salivary biomarkers (Chromogranin-A, haptoglobin), will determine if treatments affected the response to a stressful event later in life, whilst identifying the value of different biomarkers to provide information on stress.

Animal-based indicators on carcasses: a tool to monitor pig welfare and management practices on-farm and pre-slaughter

M.E. Lagoda, Y. Wang, K.A.T. Moak and Y.M. Seddon

Western College of Veterinary Medicine, University of Saskatchewan, Saskatoon, SK, S7N 5B4. E-mail: martyna.lagoda@usask.ca

Animal-based indicators (ABI) measured on carcasses are underutilized in pig welfare assessments, despite evidence of their potential to inform on welfare retrospectively. As part of a larger study investigating the value of monitoring ABI on swine carcasses as a welfare monitoring tool, this abstract reports descriptive data on the prevalence of ABI recorded from carcasses. Finisher pigs (n=3733) from 15 Western Canadian barns (26 batches, 1–2 batches/barn) were followed to slaughter at a single abattoir, where a video camera was installed to capture footage of carcass movement along the production line (post-scalding and dehairing). Recordings were observed by two trained observers and the severity of carcass ABI was scored using a 0–3 scale for skin lesions (SL; on the shoulder, middle and rump) and tail length (TL), a 0–2 scale for tail biting lesions (TBL), and for the presence or absence of loin bruises (LB) and human handling lesions (HL). Data were summarized descriptively to determine the percentage of carcasses with different ABI and severities. Moderate SL (≤10 superficial lesions) presented on 91% of carcasses, while 1% had severe SL (≥11 lesions over ≥75% of the observed area). Fresh SL reflect pig aggression during transport and lairage, and the effect of management adjustments to mitigate aggression would be rapidly tracked on carcasses. Tail-docking was evident in 99% of carcasses, representing tail-biting is a widespread concern, with 64% of tails docked short (≤2cm), likely to curb tail-biting risk. Mild TBL (small blemishes/small healing wound) presented on 40% of carcasses, while 1% had severe (open wound/necrotic tissue) TBL, showing docking alone is insufficient to curb biting. Monitoring these carcass ABI provides a feedback mechanism for additional management practices to curb tail-biting and to monitor their effectiveness. Both LB and HL were low, presenting on 8% and 2% of carcasses respectively, indicating good animal handling on-farm and pre-slaughter.

Implications: The data demonstrates the potential of monitoring carcass ABI to review management practices that work well, and those that can be adjusted to improve welfare. Carcass ABI could offer a valuable tool for providing feedback to producers and abattoir staff, for monitoring in animal care programs, and reporting animal welfare metrics for benchmarking and transparency targets.

Does the early life environment and health effect hair cortisol, dehydroepiandrosterone (DHEA) and the cortisol:DHEA ratio?

Darian Pollock[1], David Janz and Yolande Seddon

Western College of Veterinary Medicine, 52 Campus Drive, S7N 5B4, SK, Canada
[1]darian.pollock@usask.ca

Chronic stress negatively impacts swine welfare and productivity. Higher levels of the hormone cortisol, lower levels of DHEA, and a higher cortisol:DHEA ratio measured in swine hair have been proposed as biomarkers of chronic stress. Piglets undergo rapid neurodevelopment from birth to four weeks & four to twelve weeks. Experiences during this period have lifelong effects on stress responses. The goal of this study was to determine if different early life management practices influenced hair cortisol, DHEA, and the cortisol:DHEA ratio. At birth 40 litters were raised in standard systems (1 metal chain in 4.32m^2 farrowing pens) or received rope and burlap to shape oral behaviours, increased space (6m^2 farrowing pen) to shape social development, and positive human contact to reduce fear of humans. At weaning piglets (n=236 over 8 pens/batch) were either kept in the same treatment (extra space: 9pigs/pen (0.39m^2/pig vs. standard: 12pigs/pen (0.29m^2/pig)) or were switched to the opposite treatment. At 12 weeks of age all pigs were kept in standard conditions until slaughter. Hair (n=192 pigs) was collected two weeks post-weaning (reflecting pre-weaning) and at 14 weeks of age (reflecting post-weaning to 12 weeks of age) and analysed for concentrations of hair cortisol and DHEA, and their ratio calculated. Health, productivity and stress reactivity were repeatedly assessed throughout life. Treatment did not influence hair hormone concentrations or average daily gain (ADG). Piglets presenting lameness pre-weaning (n=22) had a 58% higher cortisol:DHEA ratio than healthy pigs in the farrowing period (healthy: 1.04 ± 0.12, lame:1.77 ± 0.23, p=0.004, mean ± SEM). Further analysis including the measures of individual stress variability will help determine the value of hair hormones.

Industry implications: The cortisol:DHEA ratio analyzed in hair may provide the industry with an objective biomarker of physiological stress due to lameness in piglets. The sensitivity of the hair hormones to detect management to improve welfare and stress reactivity are yet to be determined.

Burlap & buddies: Enrichment options to benefit piglets around weaning.

A. Scott[1], A. Le Heiget[1], R. Stefanson[1], J. Ahloy-Dallaire[2], and M. King[1*]

[1]Department of Animal Science, University of Manitoba, Winnipeg MB R3T 2N2.
[2]Department of Animal Science, Université Laval, Quebec, QC G1V 0A6.
*Corresponding author: Meagan.King@umanitoba.ca

Weaning is stressful for piglets as they experience many stressors at once. To mitigate piglets' stress and improve their behaviour, welfare, and performance around weaning, we compared the outcomes of two enrichment types and their combined effect in a commercial sow barn and nursery: **Object enrichment** - hanging burlap over the plastic panels in farrowing and nursery pens; and **Social enrichment** - allowing groups of 2 or 4 neighbouring litters to mix pre-weaning by removing the plastic pen dividers when piglets were ~3 days old. We recorded piglet behaviour, piglet body weights, and sow and piglet lesion scores. **By mixing piglets pre-weaning**, they can develop social skills that can be used throughout their life. Mixing piglets in groups of 4 litters pre-weaning decreased biting (P=0.03) performed by weaned piglets which improved their lesion scores 1 week post-weaning (P=0.07). Removing pen dividers provided more space per pig and more safe space, away from the sow, which may have led to fewer piglet deaths due to crushing by the sow (P=0.004). **By providing burlap**, piglets can engage in natural behaviours and have an element of familiarity between their pre- and post-weaning environments. Having access to burlap reduced biting behaviours between piglets before weaning (P=0.07) and biting and fighting behaviours after weaning (P<0.001 and (P=0.06), respectively). As a result, fewer lesions developed during (P=0.07) and after weaning (P=0.1) which may reduce disease transmission (greasy pig). Access to burlap also reduced the amount of chewing on pen objects (P=0.006), potentially improving the longevity of facilities. Overall, social (mixing) and object (burlap) enrichment is beneficial to the industry by allowing piglets to develop social skills and express natural behaviours, which does not negatively impact piglet performance and can improve piglet behaviour and welfare around weaning.

Implications: Either of these enrichment options should be considered when looking for solutions to mitigate piglet stress around weaning. They are both beneficial to piglets and can be implemented in commercial production systems.

Rearing pigs with play opportunities: the effects on disease resilience in pigs experimentally inoculated with PRRSV

K. Steinerova[1], H.L. Wilson[2], J.C.S. Harding[1], A.N. Finatto[1], S.E. Parker[1], and Y.M. Seddon[1]

[1]Western College of Veterinary Medicine, University of Saskatchewan, 52 Campus Drive, Saskatoon, SK, S7N 5B4. karolina.steinerova@usask.ca
[2]VIDO, Saskatoon, Canada; University of Saskatchewan

The association between positive emotions and improved health is established in human medicine, and initial work suggests similar benefits in farm animals. As play is linked to positive emotions, providing pigs with opportunities for play may enhance their ability to recover quickly from infectious challenge. This would help to reduce the effect of economically significant diseases such as porcine reproductive and respiratory syndrome virus (PRRSV) which negatively impactspig productivity. This study explored the effect of rearing pigs with play opportunities on disease resilience when challenged with PRRSV. Litters were assigned to either play (n=5 litters, mean: 15 pigs/litter, PLY) or control (n=4 litters, mean: 13 pigs/litter, CON) treatments at birth. In PLY, play was promoted with extra space and enrichment objects, for three hours/day from five days of age until euthanasia at 65±2 days (mean±SD). At 15 days, piglets were categorized as low, medium, or high resisters in the back test. At weaning (26±2 days), 28 pigs (14/treatment) were selected, and at 43±2 days (day 0 post-inoculation, DPI) were inoculated with 5 x 10^5 $TCID_{50}$ PRRSV/pig. Rectal temperature (RT), health scores and blood for assessment of white blood cells and viral load, were collected over a 21-day period until euthanasia (22 DPI). Viral load between the treatments did not differ. Monocytes, one of the target cells of PRRSV, decreased on 2 and 4 DPI in both groups. On 8 DPI, monocytes started to rebound but were lower in PLY than CON (p=0.005). On 21 DPI, PLY returned to baseline, while CON exceeded baseline and remained higher than PLY (P<0.001). Nine out of 14 CON pigs experienced severe respiratory distress for at least one day, compared to two PLY pigs. PLY pigs were in respiratory distress fewer days than CON (PLY:1.4±0.4, CON:5.1±1.2, (mean±SEM), p=0.004). RT was lower in high-resisting PLY pigs than in CON pigs on 8 DPI (difference:-0.5°C, p=0.003).

Implications: Play resulted in a modified response to PRRSV suggesting effects of play on the immune response. Remaining analysis will determine if this promising concept can support enhanced disease resilience in intensively farmed pigs.

The effects of periodic enrichment on the welfare and production of growing pigs

Abigail Tillotson[1,2], Yolande M. Seddon[1,2], Karen Schwean-Lardner[2], Jennifer Brown[1]

[1]Prairie Swine Centre Inc, 2105 8th Street E., Saskatoon, SK, S7H 5N9; [2]University of Saskatchewan, Saskatoon, SK, S7N 5A2.
Email: abby.tillotson@usask.ca

Environmental enrichment is known to decrease damaging behaviours and to improve pigs' ability to cope with new and stressful conditions. This study investigated the effects of a simple periodic enrichment provided in early (0-9 weeks) and/or late (9-21 weeks) life stages. Twenty-four litters were randomly assigned to two early treatments, Control (C) and Enriched (E). One week after birth, E pens were given six pieces of newsprint three days/week which was sprayed with a citrus-scented liquid (Phytozen®). At 9 weeks, half of E and C pigs moved to the opposite treatment producing four treatments in late life: EE, EC, CC, and CE (6 pens/treatment, 10 pigs/pen). Pigs were weighed at three and nine weeks and handling scores were assigned at weighing on a scale of 1 (no vocalization or struggle) to 3 (refusal to enter scale). Movement of pens from farrowing to nursery, and nursery to grow finish (GF), was timed. Skin lesions were assessed pre- and post-weaning, when moving to GF, and at 13, 17, and 19 weeks. Tail bite lesions were scored weekly from 9 to 21 weeks. All lesions were scored on a scale of 0 (no lesion) to 3 (severe lesion). Growth did not differ between treatments. At 9 weeks, E pigs had lower handling scores compared to C pigs ($p=0.014$) and faster movement from nursery to GF (E: 212±11, C: 250±11 s, $p=0.001$). At weaning, E pigs had fewer and less severe skin lesions than C pigs (mean scores, E: 0.51±1.91, C: 1.38±1.86, $p=0.004$). At 13-, 17-, and 19-weeks EE pigs had lower lesion scores than CC and EC, with CE being intermediate ($p<0.001$). Throughout GF, EE pigs had lower tail bite scores than CC and EC, with CE being intermediate ($p<0.001$).

Implications: Access to periodic enrichment resulted in fewer lesions and tail bite damage, and improved handling of pigs. The results outline the importance of maintaining enrichment throughout life, and that periodic enrichment is a practical and effective way to enrich pigs.

One-step electrocution for on-farm euthanasia of large swine: The behavioural response of sows to electrode attachment

Carmen C. Cole[1], Martine Denicourt[2], Christian Klopfenstein[3] and Yolande M. Seddon[1]

[1]Western College of Veterinary Medicine, University of Saskatchewan, 52 Campus Drive, Saskatoon, SK, S7N 5B4. E-mail: carmen.cole@usask.ca
[2]Faculté de médecine vétérinaire, Université de Montréal, Saint-Hyacinthe, QC
[3]Centre de développement du porc du Québec inc., Québec, QC

An alternative, rapid euthanasia method is needed for large pigs that is humane but more user-friendly and less dangerous for barn staff than the captive bolt gun. Captive bolt euthanasia requires stressful restraint of the pig and accuracy to target the small brain, which is difficult in sows due to a ridge bone running down the forehead. Brain damage causes bodily convulsions making the procedure visually unpleasant. One-step electrocution could provide a humane, less error-prone solution by attaching two electrical connections to the pig at locations of rich blood supply, targeting the brain and the heart between both. The procedure may also be more aesthetically acceptable; early research suggests muscle stiffening occurs during electrocution then complete relaxation without convulsions. To be safe and humane, electrodes that are easily attached and accepted by healthy pigs must be identified. Sixty-four sows received a human approach test to distinguish them as either fearful or calm to human contact. One day later, a rotation of four treatments were applied to each sow restrained in a 3m^2 pen: 1) chain bridle and belt around abdomen, 2) copper clamps on upper lip and flank, 3) copper clamps on ear and tail, 4) handling as if to apply electrodes without attachment. The treatment applied first was randomly assigned to an equal number of fearful and calm sows. Upon each treatment application, the behavioural response was scored (0=acceptance to 3=highly aversive), and electrodes removed after 10s. Up to three attempts for electrode attachment were allowed to evaluate ease of application. Results will show the electrode combination most accepted by sows and easily attached.

Implications: This investigation will ensure the electrodes used for one-step electrocution are easily applied and accepted by pigs, supporting barn staff with implementation of a simpler and more humane euthanasia method for large pigs.